DESIGN
INTELLIGENCE
AWARD

2022
中国设计智造大奖
年鉴

2022 DESIGN
INTELLIGENCE
AWARD
YEARBOOK

中国设计智造大奖组委会 编

中国建筑工业出版社

前言
PREFACE

中国设计智造大奖 (Design Intelligence Award，简称"DIA")，2014 年由李强同志提出创立，自 2016 年全球启动以来，至今已成功举办七届。回顾历程，DIA 始终积极响应时代的召唤，不断诠释着面向科技未来、设计驱动以及社会创新的独特价值观。本届大奖以"设计协同"为年度主题，涵盖"文化创新、生活智慧、产业装备、数字经济"四大参赛类别，重点关注数字赋能下社会与产业的广泛转型，挖掘与呈现人工智能、智慧出行、医疗健康、文化衍生、工业装备等多个领域的优秀范例。这些作品使我们深信，设计不仅仅是美的追求，功能性与技术的突破，更是可持续增长、民生福祉和共同价值的深度表达，其无疑推动着社会文明的发展和进步。

从本届金奖作品"盘古药物分子大模型智能加速药物研发"的趋势来看，AI 时代已经来临。AIGC 正为人类社会打开智能认知与生成的大门。我们已身处有史以来最具颠覆性的技术变革之中，这场以 AI 为代表的技术浪潮正以摩尔定律的两倍、以周或月为迭代周期飞速增长，并加快改变生产工具、生产关系以及诸多生产要素，促使整个社会生产力发生质的突破。面对这一"人类智慧"与"类人智能"互相博弈、彼此纠缠的时代，面对这场智能技术迅速消解专业门槛、激发人人平等参与创想所带来的剧变，我们提出，必须重建一种人工智能（Artificial Intelligence）与艺术智性（Artistic Intelligence）两个 AI 互相激发的，让艺术设计、智能生成、产业智造之间产生互通互融、共同进化的创新生态。

为此，本届 DIA 首度发出建立"DIU 国际设计智造联盟"的倡议。就像百年前"德意志制造联盟＋包豪斯学院"推动德国现代工业腾飞那样，这一行动，不仅意味着未来我们将以"DIA+DIU"双驱动模式，推动形成"艺术－科技－思想"充分联通、"高校－政府－企业"深度协同、"教育－产业－社会"整体创新的高质量发展新模式，凝聚形成一个融合智慧设计、智能制造，贯通人才链、创意链、产业链、传播链的国际性"创新联合体"，创建具有全球影响力的交流合作平台与跨行业网络。同时，还预示着我们将视 AI 技术为工作伙伴，让设计智慧与数字智能携手，促进文化创造与科技创新的共生共构，探索数字智能时代的未来图景，擘画"设计智造"的崭新蓝图。

本届年鉴是一部档案，更是一份宣言。它在记录过去一年中参赛者以新物种、新功能、新技术、新体验、新定义等为产业创新主战场，推动设计进化与产品迭代的点点足迹的同时，也将促发我们在新的社会关系、新的价值伦理、新的数字文明中，为实现全人类共同价值而设计，以设计智造的力量塑造更美好的未来。我们期待着，通过这份年鉴，能够为您呈现 DIA 大奖的精彩历程，激发更多关于设计、技术和创造力的深刻思考。

The Design Intelligence Award (DIA), initiated by Comrade Li Qiang in 2014, has been held for seven years since it was launched globally in 2016. Looking back to its development, DIA has always actively responded to the call of the times, and constantly interpreted the unique values of facing the future science and technology, driven by design and social innovation. Under the theme of "Design Collaboration", 2023 DIA covers four categories: "Cultural Innovation, Life Wisdom, Industrial Equipment and Digital Economy", which focus on the extensive transformation of society and industries in the context of digital empowerment, and explore and present excellent examples in fields such as artificial intelligence, smart transport, medical treatment and health care, cultural involution and industrial equipment. These works convince us that design is not only the pursuit of beauty, the breakthrough in functionality and technology, but also the in-depth expression of sustainable growth, people's livelihood and common values, and they are undoubtedly promoting the development and progress of social civilization.

"Pangu Bigmol to Accelerate Drug Discovery with AI", the 2022 DIA Gold Award works, reflected the trend that the era of AI has arrived. AIGC is opening the door of intelligent cognition and generation for human society. We are already in the midst of the most disruptive technological change in history. This technology wave represented by AI is growing rapidly at a pace twice the Moore's Law and the iteration cycle is on a weekly or monthly basis; Moreover, it is accelerating the change of production tools, production relations and multiple production factors, and promoting a qualitative breakthrough in productivity of the whole society. In the face of this era when "human intelligence" and "human-like intelligence" are playing games and entangled with each other, and in the face of the drastic changes brought about by the rapid decomposition of specialty thresholds through the intelligent technology that stimulates everyone to participate in creativity on an equal footing, we propose that we must rebuild an innovation ecology where two AIs, namely Artificial Intelligence and Artistic Intelligence, are mutually stimulated, so that art design, intelligent generation and industrial intelligence can be integrated, intercommunicated and co-evolved.

For this reason, 2022 DIA released the initiative of establishing the "Design Intelligence Union (DIU)" for the first time. Just like the "Deutscher Werkbund + Staatliches Bauhaus" that promoted the take-off of German modern industry a hundred years ago, this action means that in the future, we will, with the "DIA + DIU" dual-drive model, promote the formation of a high-quality development pattern based on the full connectivity among "art-technology-thought", deep collaboration of "university-government-enterprise" and overall innovation in "education-industry-society", and form an international "innovation consortium" integrating intelligent design, intelligent manufacturing and connecting talent chain, creativity chain, industrial chain and communication chain, thus creating an exchange and cooperation platform and a cross-industry network with international presence. In addition, the DIU initiative also indicates that we will regard AI technology as our partner, let design wisdom and digital intelligence work together to promote the symbiosis and co-construction of cultural creation and technological innovation, explore the future prospect of the digital intelligence era, and draw a brand-new blueprint of "design intelligence".

The DIA yearbook is not only an archive, but also a declaration. It records the footprints of the participants in the past year in promoting design evolution and product iteration with new species, new functions, new technologies, new experiences and new definitions as the main battlefields of industrial innovation; It will also prompt us to design for the common values of all mankind in the new social relations, new value ethics and new digital civilization, and shape a better future with the power of design intelligence. We look forward to presenting you the wonderful journey of DIA through this yearbook, and inspiring more profound thoughts on design, technology and creativity.

目录
CONTENTS

寄语
MESSAGE

汇聚工业设计创新合力，打造"工业设计高地"

中国设计智造大奖既是对浙江工业设计水平的检验，也是洞悉工业设计产业发展的年度风向标。

中共中央、国务院高度重视发展工业设计。2012 年 12 月，习近平总书记在视察广东工业设计城时强调，要提高工业设计水平，提升产品附加值，增强中国制造业竞争力。党的二十大报告指出，要推动现代服务业同先进制造业、现代农业深度融合，大力发展设计等现代服务业。前不久（2023 年 2 月 6 日），中共中央、国务院印发了《质量强国建设纲要》，提出要发挥工业设计对质量提升的牵引作用，强化研发设计，推动工业品质量迈向中高端。这些都为我们加快打造"工业设计高地"提供了重要遵循。

加快打造"工业设计高地"，是打造全球先进制造业基地、建设"415X"先进制造业集群的必然要求。工业设计，是工业价值链的起点，也是价值链竞争的制高点。我理解，设计是利用形状、材质、工艺等工具把思想转变为产品，通过创造性地解决问题而让生活更加美好。设计中的个人创意审美固然重要，但更重要的是打通艺术、技术和商业鸿沟，精准把握问题难点，深度理解客户需求，创造美好生活和更好的世界。借此机会，和大家分享三点想法和体会：

第一，浙江工业对设计有着如饥似渴的迫切需求。浙江制造业基础扎实、产业特色鲜明，拥有世界 500 强企业 9 家，民营企业 500 强 107 家（连续 24 年居全国第一）；特别是集聚了一大批专精特新中小企业，拥有国家专精特新"小巨人"企业 1067 家，重点"小巨人"200 家，单项冠军企业 189 家，数量均居全国第一位。当前，我省正在通过新一轮"腾笼换鸟、凤凰涅槃"攻坚行动，推进制造向高端化、数字化、绿色化、国际化转型升级。浙江很多企业，都曾经历过模仿、加工组装、单纯从事生产型制造的阶段，现在仍然不少企业处于价值链低端。希望广大企业家和创业者，通过中国设计智造大奖这个平台，充分感受国内外顶尖的设计力量和创意魅力，共享工业设计的赋能之美，把握产业变革趋势，推动企业由加工制造环节向合作研发、联合设计、品牌培育等全球价值链中高端持续攀升。

第二，工业设计的基本逻辑：从 ID 到 IP。工业设计（ID）的本质是沟通，通过设计更好地主动传递信息，让用户更准确方便地解码产品的信息。IP 是工业时代的品牌，反映的是价值观、世界观。从 ID 到 IP，即工业设计通过产品和服务完成沟通，也塑造了我们的生活，赋予生活更多意义的过程。有个说法，不知是否准确，第二次世界大战后，人类重建 90% 以上的现实世界。现在，人们又开始设计一个崭新的元宇宙。从这个意义上讲，没有设计就没有我们生活中的一切。这个时代里不一定每个人都成为设计师，但每个人都应该要懂设计。每个企业都应该懂得设计。

第三，融合人文之善、科学之真、艺术之美，打造"工业设计高地"。工业设计从创意到产品落地，涉及品牌策划、基础研究、专利保护、市场营销、成果落地等系列环节，是一项系统工程。要坚持以人为核心。一个好的工业设计，不只重视表面形态，而是要站在用户的角度，甚至与用户共同创造来满足用户的潜在需求。比如，进入老龄化社会后，我们希望家居用品、公共设施能够满足每一名老人独特而细微的需求。设计以人为本，这是周知的标准，却是最难的境界。要坚持工业设计与中国文化、世界元素相融合。充分挖掘中国优秀文化资源，充分发挥中国式设计文化思维优势，以大国工匠精神，更加展现文化自信、彰显中国文化底蕴，满足国民的人文精神需求。比如首届中国设计智造大奖金奖水槽洗碗机的诞生，就是典型的从中国文化与生活方式研究切入而产生的设计成果。要坚持工业设计与新材料、新技术等相结合。人工智能、生物工程等现代科技赋予未来设计无限可能。比如悠行外骨骼机器人，既注重人机交互创新设计，也强调生物科学与医疗器械技术综合创新，以及大数据、智能诊疗等统筹考虑，成为中国首款量产、可商用的下肢外骨骼医疗产品。

工业设计的蓬勃发展，有力支撑了浙江从制造大省向制造强省转变。在浙江，已经涌现出一大批设计能力十足的公司，一大批企业通过工业设计成功升级，工业设计赋予浙江制造更强的竞争力。新征程上，工业设计又迎来了新的机遇。《中共中央 国务院关于支持浙江高质量发展建设共同富裕示范区的意见》中明确要求，要打响"浙江制造"品牌，塑造产业竞争新优势。2023年初易炼红书记在"新春第一会"上也指出，当前浙江制造业总体上仍处于全球价值链中低端，要联动实施质量强省、标准强省、品牌强省战略，实现产品设计、采购、生产、销售及服务全流程协同，推动产业合作由加工制造环节向合作研发、联合设计、品牌培育等全球价值链中高端环节延伸。

七年来，这场全球设计盛宴，承载着浙江制造转型升级的坚定梦想。这里不仅有全球最顶尖的产品设计，还有最活跃的理念和思维，已成为当代创新设计评价、推广合作、吸引人才的重要平台，也是一个艺术、科技与商业的跨界创新全球竞赛，更是一个创意转向财富与未来的实体创新加速器！希望中国设计智造大奖牢记时任浙江省省长李强提出的"打造国际一流工业设计大奖"的初心，更好地赋能浙江制造业高质量发展，引领驱动新时代的设计价值。

希望大家以本次颁奖典礼为契机，碰撞思想火花、筑牢友谊基石、推进务实合作、结出丰硕成果，共同创造我们更加美好的未来。

卢山　浙江省副省长

2023 年 3 月

Aggregating Industrial Design Innovation Synergy to Create an"Industrial Design Highland"

The Design Intelligence Award serves as both an assessment of the industrial design capabilities in Zhejiang Province and an annual indicator that provides insights into the development of the industrial design industry.

The CPC Central Committee and the State Council attach great importance to the development of industrial design. In December 2012 when General Secretary Xi Jinping inspected Guangdong Industrial Design City, he emphasized that we must improve the level of industrial design, increase the added value of products and enhance the competitiveness of China's manufacturing industry. It is pointed out in the Report to the 20th National Congress of the Communist Party of China that we should promote the in-depth integration of modern service industries with advanced manufacturing and modern agriculture, and vigorously develop modern service industries such as design. Recently (February 6, 2023), the CPC Central Committee and the State Council issued the Program of Building National Strength in Quality, which proposes to give full play to the leading role of industrial design in quality improvement, strengthen R&D and design, and push the quality of industrial products towards the middle and high end. All these policies have provided important guidelines for us to promote the development of industrial design.

Speeding up the construction of "industrial design highland" is an inevitable requirement for building a global advanced manufacturing base and the 415X modern industrial cluster. Industrial design is the starting point of industrial value chain and the commanding height of value chain competition. From my perspective, design is to transform ideas into products by using tools such as shape, material and technology, and to make life better by solving problems creatively. Personal creative aesthetics in design is important, but it is more important to bridge the gap between art, technology and business, accurately grasp the difficult issues, deeply understand customer needs, and create a better life and a better world. I would like to take this opportunity to share with you three thoughts and experiences:

Firstly, industrial enterprises in Zhejiang have an urgent need for design. With a solid manufacturing foundation and distinctive industrial characteristics, Zhejiang has 9 Fortune 500 enterprises and 107 Top 500 private enterprises (ranking first in China for 24 consecutive years); In particular, the province has attracted a large number of "specialized and innovative" small and medium-sized enterprises to settle here, including 1067 national specialized and innovative "little giant" enterprises, 200 key "little giants" enterprises and 189 individual champion enterprises, ranking first in China. Now Zhejiang is promoting the transformation and upgrading of traditional manufacturing to high-end, digital, green and international manufacturing through a new round of action plans - "clearing the cage to make way for new birds". In the past, many enterprises in Zhejiang experienced such stages as imitation, processing and assembly, during which they were simply engaged in production-oriented manufacturing. Now many enterprises are still at the low end of the value chain. Taking this opportunity, I hope that entrepreneurs and start-ups can fully feel the power of international top design works and the charm of creative ideas through the platform provided by DIA, share the beauty of industrial design empowering the development of manufacturing, focus on future industrial changes, pay more attention to industrial design, and promote industrial cooperation to extend from processing and manufacturing to high-end links in the global value chain such as cooperative research and development, joint design and brand cultivation.

Secondly, the basic logic of industrial design: From ID to IP. The essence of industrial design (ID) is communication. Through design, we can deliver messages more effectively and actively, and users can decode our information more accurately and conveniently. IP is the brand in the industrial age, and it reflects the things you value and how you look at the world. The process from ID to IP, that is, the process of industrial design realizing communication through products and services, also shapes our life and gives it more meaning. Someone says - I am not sure if it is accurate - after World War II, humans have rebuilt more than 90% of the real world. Now, people are starting to design a brand-new metaverse. In this sense, everything in our life is meaningless without design. In this era, you may not become a designer, but you should know what design is. All enterprises should know what design means.

Thirdly, integrate the goodness of humanity, the truth of science and the beauty of art to create a "highland of industrial design". From creativity to product realization, industrial design, as a system engineering, involves multiple links such brand planning, basic research, patent protection, marketing and achievement application. So we should center on meeting people's demands. A good industrial design not only pays attention to the superficial form, but stands on the user's point of view, and even creates together with the users so as to meet their potential needs. For example, in an ageing society, we hope that household supplies and public facilities can meet the unique and subtle needs of every elderly person. Though people-oriented design is a well-known standard, it is the most difficult state to achieve. We should integrate industrial design with Chinese culture and international elements. We should give full play to China's excellent cultural resources, and the advantages of Chinese design culture and thinking, and show our confidence in Chinese culture, highlight Chinese cultural deposits and meet the intellectual and cultural needs of the people with the spirit of craftsmen as a great power. For example, the sink dishwasher, Gold Award winner of the first DIA, is a typical design achievement based on the research on Chinese culture and lifestyle. We should combine industrial design with new materials and technologies. Modern technologies such as artificial intelligence and bioengineering give infinite possibilities for future design. For example, the UGO Exoskeleton Robot not only pays attention to the innovative design of human-computer interaction, but also emphasizes the comprehensive innovation of biological science and medical device technology, as well as the overall consideration of big data and intelligent diagnosis and treatment, so that it becomes the first mass-produced and commercially available lower limb exoskeleton medical product in China.

The vigorous development of industrial design has strongly supported Zhejiang's transformation from a big manufacturing province to a strong manufacturing province. A large number of companies with strong design capacity have emerged in Zhejiang, and many enterprises have been upgraded through industrial design. Industrial design gives Zhejiang manufacturing industry stronger competitiveness. On the new journey, we are ushering in new opportunities of industrial design. The Opinions of the CPC Central Committee and the State Council on Supporting High-quality Development and Construction of Demonstration Zones for Achieving Common Prosperity in Zhejiang clearly requires that Zhejiang should cultivate the brand of "Made in Zhejiang" to create new advantages in industrial competition. At the beginning of the year, Mr. Yi Lianhong, Secretary of Zhejiang Provincial Party Committee, also pointed out at the "First Meeting of the New Year" that Zhejiang's manufacturing industry is still at the low end of the global value chain as a whole. So we should jointly implement the strategy of building Zhejiang Province with powerful quality, standards and brands, realize the whole-process collaboration of product design, procurement, production, sales and service, and promote industrial cooperation to extend from processing and manufacturing to high-end links in the global value chain such as cooperative research and development, joint design and brand cultivation.

In the past seven years, this global design feast has carried the unswerving dream of Zhejiang's manufacturing transformation and upgrading. Here we have the world's top product designs, and the most active ideas and thinking. DIA has become an important platform where we evaluate and promote contemporary innovative design works, carry out cooperation and attract design talents, a global competition for cross-border innovation in art, technology and business, as well as an innovation accelerator for entities to turn their creativity to wealth and future. I hope that DIA can keep in mind the original aspiration of "building a world-class industrial design award" put forward by Mr. Li Qiang, the then governor of Zhejiang Province, so as to better empower the high-quality development of Zhejiang manufacturing industry and lead the design value that drives the new era.

I also hope that we will take this Award Ceremony as an opportunity to spark new ideas, build a solid foundation of friendship, promote pragmatic cooperation and yield fruitful results. Let's work together to create a better future.

Lu Shan, Vice Governor of Zhejiang Province

March , 2023

从 DIA 到 DIU

DIA（中国设计智造大奖）2014 年由李强同志倡议，2015 年正式创建，至今已是第八个年头。八年来，我们赢得了世界设计师组织（WDO）等百余家国际设计机构的鼎力支持，在全球范围内征集了 70 多个国家和地区的参赛作品近 43000 余件，连动全球 4200 家企业、近 800 所院校和 20000 多名设计师，为数字经济、智能社会的创新发展建立起了一个全球性的设计产业智库。八年来，DIA 以专业性、国际性、前瞻性为旨归，逐渐发展成为中国最具世界影响力的工业设计赛事，更成为全球设计界探讨智能设计与设计智慧的重要平台。

2020~2022 年，在全球各地建立了 13 个国际工作站、12 个海外分赛区，展现了 DIA 开放包容的学术胸襟，促进了全球设计界的国际团结。我们要向大奖团队表示衷心的感谢！饮水思源，我们更要感谢中国设计智造大奖的发起人李强同志，正是他的远见卓识，为我们创建了这个联通中国与世界、融合艺术与科技、协同设计与产业的国际学术平台。

当下，新时代征程、新发展格局正全面展开，我们迎来了新的发展机遇，DIA 将以更高的站位、更大的抱负再次扬帆启航，在"艺创＋科创"的蓝海探索、遨游。

一百多年前，德国设计界、教育界、工业界的一批有识之士先后创建了"德意志制造业同盟"和包豪斯学院，将工作室中艺术家的手与生产线上劳动者的手链接在一起，实现了从理念、设计到生产的全流程升级，造就了德国制造业的黄金时代，更创建了现代工业设计的伟大开端。今天，中国美术学院愿携手 DIA 所连带的所有艺术院校、设计师组织、数字科技与先进制造企业，共同倡议，建立一个高能级的创新联合体——"国际设计智造联盟"，我们称之为"DIU"。

今天的典礼精英荟萃，嘉宾云集，一会儿我们将正式发布倡议，共同发起 DIU。我们将以"长三角"为基地，以开放共享之精神，让人才、科技、资金、创意、知识贯通流动起来，为智慧设计与智能制造构建最优质的创新生态。

我们身处一个大变革的时代，科技迭代、大国博弈正将我们带向不确定的未来。让我们抛开一切狭隘和割裂，超越所有纷争与鸿沟，以艺术的诚意、设计的善意，拥抱最新的科技、最前沿的产业、最挑战的知识、最遥远的他者，让我们汇聚艺术设计、智能科技、数字经济的力量，打造一个多方协同、多元开放的"创新联合体"，一个艺术、科技、产业共生共创的未来设计的"反应堆"。

浙江是全球智能科技和数字经济发展的重要现场。2003 年，时任浙江省委书记的习近平同志以极具前瞻性的战略眼光，提出"数字浙江"建设方略。二十年后的今天，浙江已经成为智能制造、电子商务、金融科技的最前线，正在加速成为世界性的网络科技和创新经济重镇。我们相信，在浙江省政府的大力支持下，各界精英人士通力协作，全球设计师热情参与，DIA 和 DIU 一定会成为产业更新的助推器、区域发展的加速器，一定会成为全球制造业迭代的关键力量。我们脚下这片新时代的热土，也必将成为数字科技、智能设计、混合现实未来场景的实验场，成为数字经济、智能制造、行业创新的策源地。

DIA 再起航，DIU 就在前方。我们期待着 DIU 为我们链接起更多的同道、更丰富的资源，我们更期待着 DIA 不断为我们带来最新的探索、未来的消息！

高世名 中国美术学院院长

2023 年 3 月

DIA GOES TO DIU

The eight-year-old DIA was initiated by Mr. Li Qiang in 2014 and formally established in 2015. In the past eight years, we have got strong support from more than 100 international design institutions including the World Design Organization (WDO). We have received nearly 43,000 entries from more than 70 countries and regions worldwide, which connect 4,200 enterprises, nearly 800 colleges and more than 20,000 designers around the world to establish a global design industry think tank for the innovative development of digital economy and intelligent society. Over the eight years since its establishment, DIA has gradually developed into the most influential industrial design competition in China with the aim of professionalism, internationality and foresight, and has become an important platform for the global design community to discuss intelligent design and design wisdom.

In 2020-2022, the DIA team has overcome various difficulties and established 13 international workstations and 12 over-seas divisions around the world, which not only demonstrates DIA's open and inclusive academic mind, but also promotes the international unity of the global design community. We would like to express our sincere appreciation to the DIA team! As an old Chinese saying goes, "When drinking water, one should think of its source". We would like to thank Mr. Li Qiang, the initiator of DIA. It is his foresight that has led to the birth of this international academic platform that connects China with the world, integrates art and technology, and aligns design with industry.

Nowadays, the journey in the new era and the new development pattern are in full swing, and we are ushering in new development opportunities. DIA will set sail again with a higher position and greater ambition, and explore and roam in the blue ocean of "Artistic Innovation + Technological Creation".

More than a hundred years ago, a group of men of insight in the design, education and industry circles in Germany successively established Deutscher Werkbund and Staatliches Bauhaus, which linked the hands of artists in the studio with those of workers in the production line, upgraded the whole process from concept, design to production, and thus created the golden age of the German manufacturing industry and the great beginning of modern industrial design. Today, CAA, together with all art colleges, designer organizations, digital technology and advanced manufacturing enterprises connected by DIA, proposes to establish a high-level innovation consortium - the "Design Intelligence Union", or as we call it, "DIU".

Today, the ceremony is full of elites and guests. In a moment, we will officially announce our initiative and jointly launch the DIU. Based on the Yangtze River Delta and in the spirit of openness and sharing, we will push talents, technology, funds, creativity and knowledge to flow, and create the best innovation ecology for intelligent design and intelligent manufacturing.

We are in an era of great changes when the iteration of technologies and the super power game are bringing us to an uncertain future. Let's put aside narrowness and separation, go beyond all disputes and gaps, embrace the latest technology, the most cutting-edge industry, the most challenging knowledge and the otherness in the distance with the sincerity of art and the goodwill of design. Let's gather the strength of art design, intelligent technology and digital economy to create a collaborative and diverse "innovation consortium", a "reactor" for future design with the co-existence and co-creation of art, technology and industry.

Zhejiang is an important site for the development of global intelligent technology and digital economy. In 2003, Mr. Xi Jinping, the then secretary of Zhejiang Provincial Party Committee, put forward the construction strategy of "Digital Zhejiang" with a forward-looking strategic vision. Twenty years later, Zhejiang has become the forefront of intelligent manufacturing, e-commerce and financial technology, and is accelerating to become an international center of network technology and innovative economy. We believe that with the strong support of Zhejiang provincial government, the cooperation of elites from all walks of life and the enthusiastic participation of global designers, DIA and DIU will surely become the boosters for industrial transformation, accelerators for regional development and key forces for global manufacturing iteration. In the new era, the land under our feet will also become an experimental field for digital technology, intelligent design and the mixed scenario combining the present and the future, as well as the source of innovation in digital economy and industry.

Now DIA is setting sail again, and DIU is waiting for us in the near future. We hope that DIU can attract more like-minded friends and more abundant resources; we are also expecting that DIA will constantly bring us the latest discoveries and futuristic messages!

Gao Shiming, President of China Academy of Art

March , 2023

设计需要一点灵性，要有"心有灵犀一点通"的感觉，要突破现有的设计，突破自己现有的边界；其次，设计要有让人有动心的感觉，然后再来追求一个境界；最后就是以总的、更高的视角去看待设计，需要有一个更大的格局。

Design requires a touch of spirit, to arouse a profound connection and tacit understanding. The constraints of the existing design should be broken to push the boundaries forward. Moreover, design should be able to evoke a feeling of fascination before fulfilling the aspiration for a higher realm. Lastly, the perception of design should be based on a holistic and broader perspective, which requires a big picture view.

宋建明 Song Jianming

中国 China
中国美术学院教授，中国智造大奖 DIA 组委会主席
Professor of China Academy of Art, Chairman of the
Organizing Committee of DIA, China Smart Awards

设计总还是设计，是去构想、去规划，去解决问题、去创造新的可能。但设计的对象、材料和环境在发生不断的演化。设计的对象更加宽泛，从器物到系统到服务；设计的材料也从自然材料、合成材料向数字和信息材料拓展；设计所面对的系统生态和社会生态也更加复杂且互相交织、互相依赖。设计会面临更多、更广泛的机遇和挑战。

Design, at its core, is about the act of designing itself, about to conceptualize, to plan, to solve problems and to create new possibilities. However, the objects, materials, and environments of design are constantly evolving. The design objects become broader in scope, ranging from artifacts to systems and to services, and the materials expand from natural and synthetic materials to digital and information materials. Additionally, the system and social ecology in front of design have also become more complex, as well as intertwined and interdependent. Therefore, design will be faced with more and wider range of opportunities and challenges in the future.

胡军 Hu Jun

中国 China
荷兰埃因霍温科技大学工业设计学院副教授，欧中设计促进会主席
Associate Professor at the School of Industrial Design, Eindhoven
University of Technology,Dutch, and Chairman of the Europe-China Design
Promotion Association

现在，设计比以往任何时候都更能发现自己处于人类努力的几乎所有领域的汇合点。设计师一直是技术和人类价值之间以及文化和商业之间的调解人。但是现在，我们的梦想能力和我们的行动能力之间的差距正在缩小，我们的身体和我们的思想之间的分界线也在缩小。作为设计师，我们现在面临的挑战是让人们看到性能和思想的优雅，同时扩大我们对有关技术、健康和环境的全球对话的责任。

Now more than ever, design has found itself at the convergence of virtually all areas of human endeavor. Designers have always been mediators between technology and human values and between culture and commerce. But now that the gap between our ability to dream and our capacity to do is narrowing, so too is the dividing line between our corporeal existence and our intellectual one. Our challenge as designers now is to make visible the elegance of performance and thought while simultaneously expanding our responsibility to the global dialogue regarding technology, health and the environment.

乔什·欧文 Josh Owen

美国 America
罗切斯特理工大学维涅利设计研究中心主任、特聘教授
Director and Distinguished Professor, Vignelli Center for Design Research,
Rochester Institute of Technology

通过在以用户为中心的设计解决方案中实施最新技术，未来的设计将成为解决我们在能源、食品和生态方面面临巨大挑战的主要工具。设计思维已经处于所有行业创造性解决问题的最前沿，并将成为创造颠覆性创新以提供综合解决方案的主要工具。设计师将在任何组织中发挥关键作用。

By implementing the latest technologies in user-centered design solutions, the design of the future will be the main tool for solving the great challenges we face in energy, food and ecology. Design thinking is already at the forefront of creative problem solving across all industries and will be the primary tool for creating disruptive innovations to deliver integrated solutions. Designers will play a key role in any organization.

马西斯·海勒 Mathis Heller

德国 German
德稻汽车及工业设计大师、中国工业设计协会 - 德稻创新学院副院长
Master of automotive and industrial design of Tokudo, Vice President of
China Industrial Design Association - Tokudo Innovation Academy

未来将以我们今天甚至无法想象的方式融合人工智能，特别是在医疗、健康和教育领域。最初是一种有趣的娱乐技术，将演变成一种融入产品和服务的功能。设计面临的挑战是确保新技术不会取代人类交互，而只是使其更好、更快或更容易使用。只有当设计控制并管理用户体验以增强人性时，这才会发生。

The future will incorporate AI in ways we cannot today even imagine—especially in the areas of medical, health, and education. What began as a fun entertaining technology will morph into a feature incorporated into products and services. The challenge for design is to ensure that new technology doesn't replace human interaction but just makes it better, faster, or easier to use. That will only happen when design take control and manages the user experience to enhance humanity.

劳拉·德斯·恩方斯 Laura Des Enfants

美国 America
Core77 咨询顾问
Consultant for Core77

1. 设计将继续成为一种战略力量，将创意产业和传统产业结合起来。这种力量以我们的星球和人类为中心，将进一步发展它所产生的效益，它将创造新的视野和文化。设计，是创新文化的最重要的面孔之一，将继续成为保存其存在的最重要的职业之一，并在未来的时期内有更多的责任。

2. 在人机关系和我们的生活中，IQ、EQ 和 II（直觉智能）的存在将继续向有利于机器的方向发展。我希望我们人类，特别是 EQ 和 II，将拥有它们的存在，我们可以保持这些感觉。更加智能的设备，以不同的智能组合更多地参与到我们的生活中，它们将以其在几乎每个领域的资产和性能开始一个复杂的旅程。

1. Design will continue to be a strategic power that brings creative industries and conventional industries together. This power, which takes our planet and humanity to its center, will further develop the benefits it produces with the new vision and culture it will create. Design, which is one of the most important faces of innovative culture, will continue to be one of the most important professions that preserves its existence and has increased responsibilities in the coming period.

2. The presence of IQ, EQ and II (Intuitive intelligence) in the human-machine relationship and in our lives will continue to develop in favor of machines. I hope that we humans, especially EQ and II, will own their existence and we can keep these feelings. Smarter devices, which are more involved in our lives with different combinations of intelligence, will have started a sophisticated journey with their assets and performances in almost every sector.

塞尔塔克·埃尔赛因 Sertac Ersayin

土耳其 TURKEY
WDO 董事会成员，土耳其工业设计师协会副主席
Member of the Board of Directors of WDO and Vice President of the Industrial
Designers Association of Turkey

设计是在真实可行的项目中转化视觉的文化和方法。设计不仅仅是产品或服务的设计，它越来越成为一种系统设计，以一种整体的方法始终考虑自然和社会环境。设计将越来越多地成为协作链的"主角"，从思考、创造和研究开始，通过过程和技术，传播和物质或非物质产品和服务的分配。设计将越来越成为与全球场景相关的数量和质量之间的"平衡者"。设计必须合作研究生活和经济的新模式，专注于多种人际关系质量的创新：人与人，人与社会，人与自然……因此，设计将变得越来越具有战略性和复杂性。

Design is the culture and method to transform vision in real and feasible projects. Design is not only products or services design but more and more became a System Design, with a holistic approach always considering the natural and social environments. Design will become more and more a "catalyst leading actor" into collaborative chains, starting from thinking, creativity and researching, through processes and technologies, communication and distribution of material or immaterial products and services. Design will become more and more a "balance maker" between quantity and quality related to the global scenarios. Design has to collaborate in the research of new models in life and economy focused on the innovation of the quality of the multi human relationships: human to human, human to society, human to nature...So, Design will become more and more strategic and complex.

卢卡·福伊斯 Luca Fois

意大利 Italy
米兰理工大学设计学院教授、创意咨询专家
Creative Consultant for Professors at the School of Design, Politecnico di Milano

无论人工智能的发展达到什么样的高度，"人"的参与和干预，还是不可或缺。人对于充分利用发挥机器智能，挖掘大数据创造价值，也扮演至关重要的角色。例如，数据可视化，将人类知识纳入基于数据的洞察，以及决策支持系统。设计者对未来人机关系的深度思考非常重要，也会影响深远，因为人机关系将塑造未来专业人员和管理人员的新角色和知识体系。

Regardless of how advanced is the development of artificial intelligence (AI), human involvement and intervention remain essential. Meanwhile, humans also play a crucial role in fully harnessing and exploring machine intelligence, as well as creating value from big data. For instance, data visualization incorporates human knowledge into data-based insights and decision support systems. It is crucial and far-reaching for designers to have in-depth thinking about the future human-machine relationship, as this relationship will form new roles and knowledge systems for professionals and managers in the future.

宗福季 Fugee TSUNG

中国香港 Hong Kong,China
香港科技大学（广州）工业信息与智能研究所所长，香港科技大学讲席教授
Director, Institute of Industrial Information and Intelligence, The Hong Kong
University of Science and Technology (Guangzhou), and Chair Professor, The Hong
Kong University of Science and Technology

获奖作品
AWARD-WINNING WORKS

金奖
2022 DIA
GOLD
AWARD

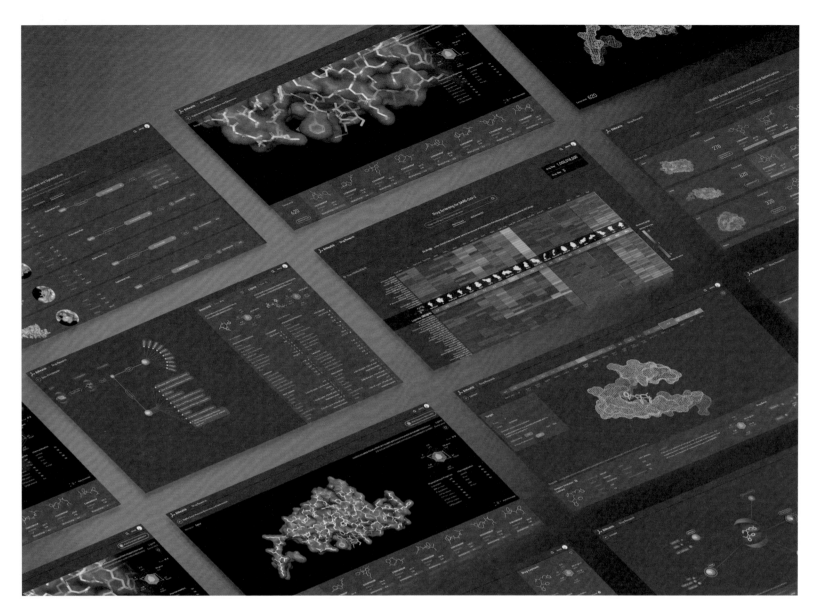

盘古药物分子大模型智能加速药物研发
PANGU BIGMOL TO ACCELERATE DRUG DISCOVERY WITH AI

盘古药物分子大模型是一款通过 AI 技术辅助科研人员加速药物研发的平台。华为云盘古药物分子大模型通过 AI 技术辅助科研人员加速药物研发。药物研发全流程可视可感知，借助云上大算力和 AI 模型，海量药物分子数据筛选和药物分子优化生成的速度得到极大提升，大幅度缩短了传统药物研发的试错过程。华为云联合西交大第一附属医院研发出全新光谱抗菌药，研发周期从数年缩短至一个月。

Pangu Drug Molecule Model is an AI R&D platform to accelerate drug discovery. HUAWEI Cloud Pangu Drug Molecule Model is a pretrained AI model to accelerate drug discovery. The whole process of drug discovery is visible and perceptible. With cloud computing power and AI models, the speed of massive drug molecules screening and drug molecules optimization has been greatly accelerated. Pangu drug molecule model greatly shortens the trail-and-error process of traditional pharmaceutical industry. Released new broad-spectrum antibacterial drugs jointly developed by the First Affiliated Hospital of Xi'an Jiaotong University, the R&D cycle was shortened from years to one month.

作者：许冰华 李真珍 王咿临 乔楠 熊招平
机构：华为技术有限公司 / 华为技术有限公司2012实验室 UCD 中心
国家：中国
组别：产业组

AUTHOR : Xu Binghua, Li Zhenzhen, Wang Yilin, Qiao Nan, Xiong Zhaoping
UNIT : Huawei Technologies Co., Ltd. / Huawei Technologies Co., Ltd. 2012 Laboratories UCD Center
COUNTRY: China
GROUP : Product Group

专家点评

这是一款帮助人类简化技能要求、降低试错成本、缩短探索路径、拓展知识边界的药物科技平台。它的出现让药物研发全流程可感知，更友好的用户体验和易用性能辅助科研人员更高效地完成智能药物筛查和药物优化生成，积极推动了传统制药行业向 AI 技术辅助药物设计的转型升级。在实验室和患者应用之间的鸿沟上架起了一座云梯。设计者用云上大算力和 AI 模型的双重加持让药物研究的每一步跋涉不再全凭运气，而是有了坚实的后盾。

EXPERT REVIEWS

It is a pharmaceutical technology platform that helps humans simplify skill requirements, reduce trial and error costs, shorten exploration paths and broaden the boundaries of knowledge.Its emergence makes the whole process of drug R&D perceptible. Its friendlier user experience and ease of use can assist researchers completing intelligent drug screening and optimized drug generation more efficiently, and actively promote the transformation and upgrading of the traditional pharmaceutical industry to the AI-assisted drug design. It bridges the gap between laboratory and patient applications. With the dual support of cloud computing power and AI model, designers make every step of drug research no longer a matter of luck, but under solid backup.

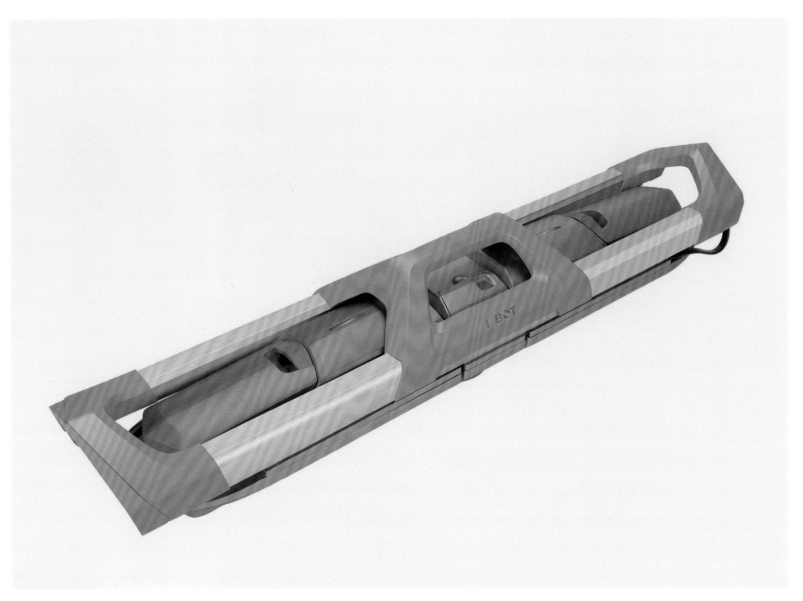

便携式智能光伏清扫机器人
IFBOT X3 SOLAR PANEL CLEANING ROBOT

打破传统观念束缚，重新定义便携式光伏清扫机器人。机器人体积小巧重量轻，无水清洁，智能环保。不同于传统的作业模式，翼博特光伏清扫机器人采用轻量化与智能化设计理念，在数公斤的单机重量内，配备了完全自主清洁、路线规划、低电返航等功能。清洁无需耗水即可获得最佳的清洁效果，并具备超强吸尘能力与壁面角度适应性。超轻的重量更可减少使用者的搬运负担以及对脆弱光伏面板的伤害。

A Brand New way to clean solar panels. IFBOT X3 Cleaning Robot's design of lightweight and intelligence make solar cleaning more efficient and humanized. It adopts water-free cleaning which is more environmentally-friendly. Traditional solar cleaning often has the drawback of high equipment and labor costs. To solve this, IFBOT TECH take the lead in launching the IFBOT X3 Solar Panel Cleaning Robot. The features of lightweight and intelligence make solar cleaning more efficient and humanized. IFBOT cleaning system integrates AI technologies of fully autonomous cleaning, substantial dust removal, route planning, low-power returning and adopts water-free cleaning with a max gradeability of 45 degrees. The robot's small size, powerful function, and portable design allow it to be easily carried to some special workplaces such as steep mountains and rooftop.

作者：沈象波 卢新城 吕知轩 陈亮 花朋
机构：苏州翼博特智能科技有限公司 / 苏州轩昂工业设计有限公司
国家：中国
组别：产业组

AUTHOR : Shen Xiangbo, Lu Xincheng, Lv Zhixuan, Chen Liang, Hua Peng
UNIT : Suzhou Ifbot Intelligent Technology Co., Ltd. / SuperDesign (Suzhou) Co., Ltd.
COUNTRY : China
GROUP : Product Group

专家点评

相比风电、水电和核电，光伏发电是目前所有能源品种中促进绿色转型的最佳选择，是未来三十年世界实现碳中和目标的希望之光。这款产品的设计诞生，将广泛造福我国西北风沙地区的大型光伏发电站，解决传统清洗设备力所不逮的现状。它也让我们看到，哪怕在最偏远的角落，设计也能改变生活的现状，它们是先行者，更是理想家。

EXPERT REVIEWS

Compared with wind power, hydropower and nuclear power, photovoltaic power is the best choice among all types of energy for promoting green transformation, and is the hope for the world to achieve carbon neutrality in the next 30 years. This product will widely benefit the large photovoltaic power stations in the windy and sandy areas of Northwest China and solve the shortcoming of traditional cleaning equipment. It also shows us that design can change the reality of life even at the most remote corners. It is a pioneer and an idealist.

银奖
2022 DIA
SILVER
AWARD

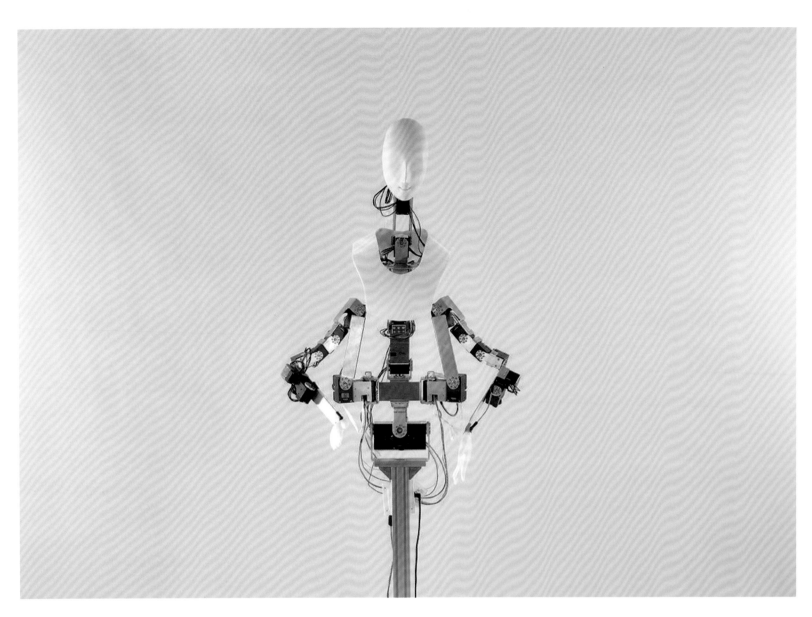

BUNRAKU PUPPET ROBOT BR-02

文乐木偶机器人 BR-02具有骨骼伸缩等新结构，设计为可进行传统仿人机器人所不具备的情感和类似生物的动作。这些成果可以提高人类与机器人之间的沟通和亲和力，并为人类和机器人创造最佳的共生关系。

The Bunraku Puppet Robot BR-02 has a new structure such as skeletal expansion and contraction, and is designed for emotional and creature-like movements not found in conventional mechanical humanoid robots. These achievements can improve communication and affinity between humans and robots, and create optimal symbiosis for humans and robots.

作者：Shinobu Nakagawa, KAKENHI (A)16H01804 Research
Members and Others
机构：Osaka University of Arts
国家：日本
组别：产业组

AUTHOR : Shinobu Nakagawa, KAKENHI (A)16H01804
Research Members and Others
UNIT : Osaka University of Arts
COUNTRY : Japan
GROUP : Product Group

专家点评

文乐木偶机器人 BR-02（Bunraku Puppet Robot BR-02）以日本延续400多年的传统艺术"文乐木偶"为灵感，通过学习其骨骼伸缩等动态特征，赋予了传统人形机器人所没有的情感表现及生物动作，创造了人与机器、科技与文化的共生关系，并为传统文化注入新的生命和活力。

EXPERT REVIEWS

Bunraku Puppet Robot BR-02 is inspired by Bunraku Puppet, a traditional Japanese art that has been passed on for more than 400 years. By learning its dynamic features such as bone expansion and contraction, Bunraku Puppet Robot BR-02 has the emotional expression and biological action that traditional humanoid robots do not have. It has created a symbiotic relationship between man and machine, and technology and culture, and injected a new life and vitality into traditional culture.

HiPhi Z

高合是新世界的高端豪华品牌，HiPhi Z 作为高合品牌双旗舰之一，定位高端豪华电动车市场。高合 HiPhi Z 充分实践"场景定义设计"理念，业内首创星环光幕 ISD 系统将灯光智能交互推向新的高度，外观独特而震撼的比例创造了外观的显著未来感。HiPhi Z 使用 NT 电磁对开门，为用户营造极具仪式感的智能进出方式。

HiPhi is a high-end luxury brand in the New World. As one of the dual flagships of the HiPhi brand, HiPhi Z is positioned in the high-end luxury electric vehicle market. HiPhi Z fully implements the concept of "scenario defines design". The industry's first wraparound star ring ISD light curtain system pushes the intelligent interaction of lighting to a new height. The unique and shocking proportions of the exterior create a significant futuristic appearance. HiPhi Z uses NT electromagnetic suicide door to create a smart way of entering and exiting for users with a sense of ritual.

作者：尼古拉斯 赫英 尼尔斯 李颖红 卡梅拉
机构：华人运通
国家：中国
组别：产业组

AUTHOR : Nicolas HUET, Martin He, Nils Uellendahl, Sandy
Li, Camilla Kropp
UNIT : Human Horizons
COUNTRY : China
GROUP : Product Group

专家点评

这是一款将"场景定义设计"理念做到极致的电动车产品。业内首创星环光幕 ISD 系统、NT 电磁对开门，为用户营造极具仪式感的灯光和进出体验。双调性数字座舱、多轴位移机械人创造的声光电交互效果，为前后排乘客营造了别具一格的行驶氛围。使用者将在这款设计上，获得充分的沉浸式驾驶体验。

EXPERT REVIEWS

It is an electric vehicle product that maximizes the concept of "scenario defines design". It adopts the industry's first wraparound star ring ISD light curtain system and NT electromagnetic suicide door to create a smart way of entering and exiting for users with a sense of ritual. The sound, light and electricity interaction effect generated by the dual-tone digital cockpit and multi-axis displacement robot forms a unique driving atmosphere for the front-seat and back-seat passengers. From this design, users will enjoy a fully immersive driving experience.

DARWIN BUCKY

Darwin Bucky 是一种多功能预制参数化建筑"外骨骼",它是一种"易于插入"的模块化轻型独立(或集群)结构,挑战了传统建筑的固有固定性;在任何地点对场地的影响都最小;可在几天内组装或拆卸;可在集装箱中储存/运输。

Darwin Bucky is a multifunctional prefabricated parametric exoskeleton that challenges the intrinsic immobility of conventional architecture, by being a modular lightweight stand-alone (or clustered) structure that is"easy-to-plug-in"; with minimal site impact at any location; assembled or dismantled in a few days'time; stored/transported in a shipping container.

作者：Abhay Mangaldas, Jwalant Mahadevwala
机构：Darwin Projects, Andblack Design Studio
国家：印度
组别：产业组

AUTHOR：Abhay Mangaldas, Jwalant Mahadevwala
UNIT：Darwin Projects, Andblack Design Studio
COUNTRY：India
GROUP：Product Group

专家点评

Darwin Bucky 是采用预制化零件，可在现场进行快速组装和拆卸的建筑设计。通过参数化的设计，它能适用于多种场合活动及反复被使用。它的出现重新定义了人与建筑的关系，"我们是这个星球的临时访客，不能留下任何影响子孙后代的足迹"。

EXPERT REVIEWS

Darwin Bucky is an architectural design using prefabricated parts that can be quickly assembled and disassembled on site. Through parametric design, it is applicable to various occasions and activities and can be used repeatedly. Its emergence redefines the relationship between humans and buildings. "We are temporary visitors to this planet and shouldn't leave any footprints that will affect future generations."

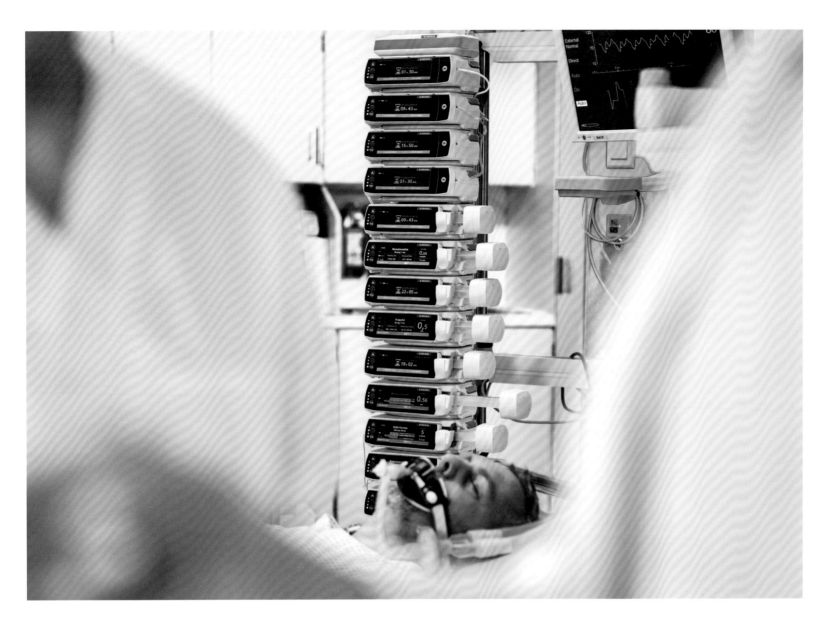

SPACEPLUS INFUSION PUMP SYSTEM

B.Braun Space Plus 系统设计用于重症监护病房（ICU），包括两种泵类型：注射泵（Perfusor®）和大容量泵（Infusomat®），以及一个扩展坞。该系统以其紧凑、便携和坚固的设计彻底改变了重症监护室的工作场所。它具有集成手柄、耐用电池和易于维护的灵活性。每个泵都可通过 WiFi 连接，将重要的治疗数据传输到医院系统。设计以可靠性、安全性和可用性为优先考虑，注重功能和用户需求。

The B. Braun Space plus System is designed for the use in Intensive Care Units, including 2 pump types, syringe pump (Perfusor®) and large volume pump (Infusomat®), as well as a docking station.The System revolutionizes ICU workplaces with its compact, portable, and robust design. It offers flexibility with integrated handles, durable batteries, and easy maintenance. Each pump connects via WiFi to transmit vital therapy data to the hospital's systems. Prioritizing reliability, safety, and usability, the design focuses on function and user needs.

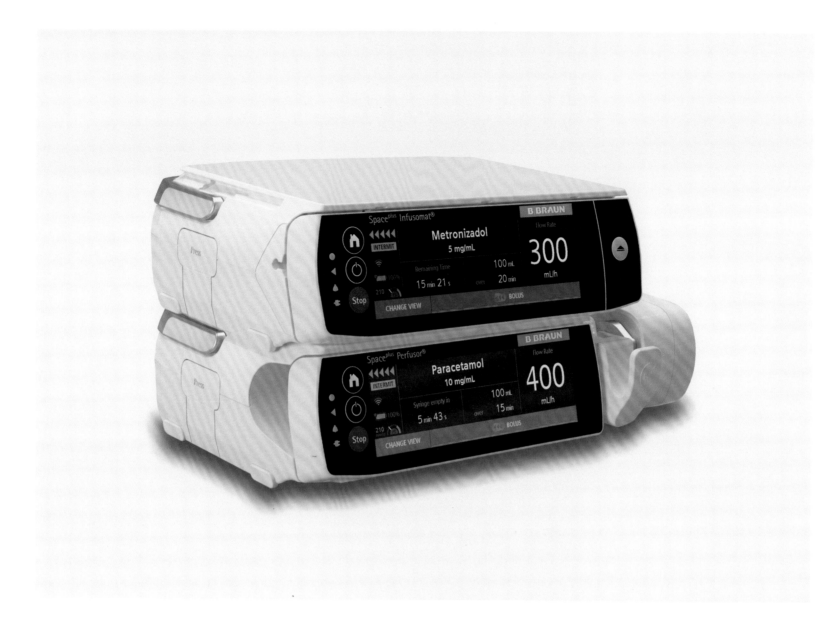

作者：Norbert Koop, Stefan Espenhahn, Jan Sokoll
机构：B.Braun Melsungen AG
国家：德国
组别：产业组

AUTHOR : Norbert Koop, Stefan Espenhahn, Jan Sokoll
UNIT : B.Braun Melsungen AG
COUNTRY : German
GROUP : Product Group

专家点评

Spaceplus Infusion Pump System 是专为重症监护病房设计的数字输液系统。模块化的设计使它既能满足特殊环境下的空间限制又具有扩展性，还可通过数字化系统进行实时管理和监控，为此类工作场所的设计树立了行业标杆。

EXPERT REVIEWS

The Spaceplus Infusion Pump System is a digital infusion system designed exclusively for ICUs. The modular design enables it to meet the requirements for space constraints and scalability in special environments, and to be managed and monitored in real time through a digital system, setting an industry benchmark for the design of such workplace.

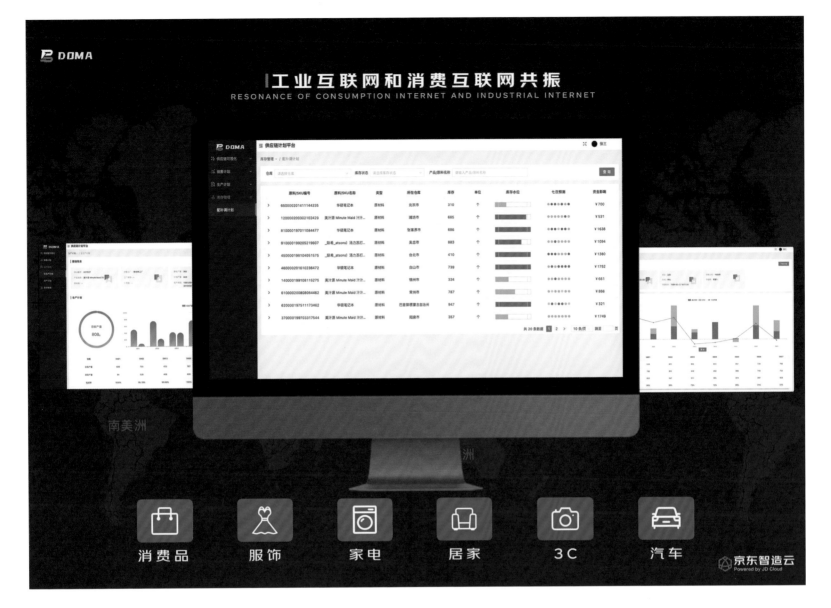

京东智造云新品孵化平台
NEW PRODUCT INCUBATION PLATFORM OF JD ZHIZAOYUN

新品孵化平台是致力于帮助生产制造企业实现产品趋势洞察、产品工业设计、智能仿真测款、产品运营等服务的智能平台。可将新品设计周期从3~12个月降到1~2个月，可实现5亿＋用户的精准触达，使新品上市成功率有了极大提高。

The new product incubation platform is an intelligent platform dedicated to helping manufacturing enterprises realize product trend insight, product industrial design, intelligent simulation testing, product operation and other services. It can reduce the new product design cycle from 3-12 months to1-2 months and achieve accurate touch of more than 500 million users. Therefore, the success rate of new products is greatly improved.

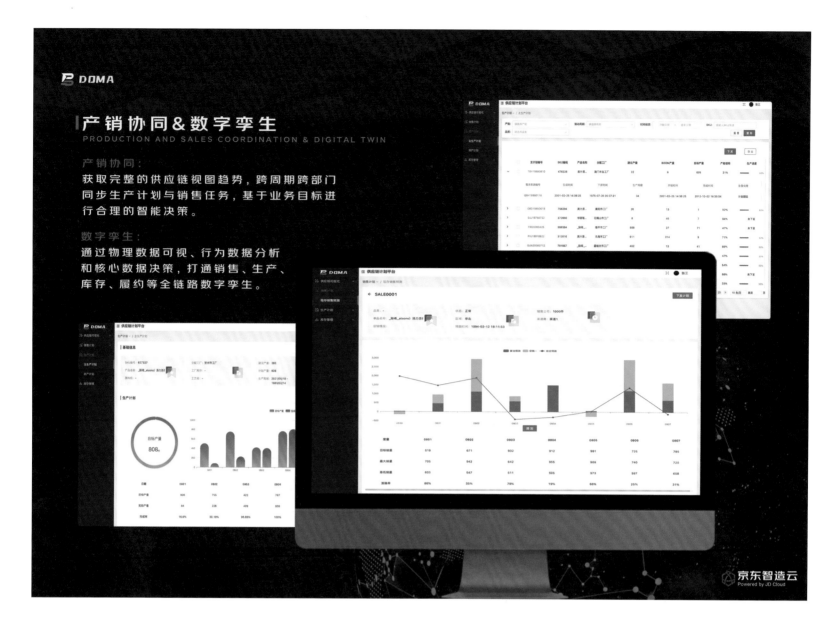

产销协同 & 数字孪生
PRODUCTION AND SALES COORDINATION & DIGITAL TWIN

产销协同：
获取完整的供应链视图趋势，跨周期跨部门
同步生产计划与销售任务，基于业务目标进
行合理的智能决策。

数字孪生：
通过物理数据可视、行为数据分析
和核心数据决策，打通销售、生产、
库存、履约等全链路数字孪生。

作者：许舟平 张璨 黄嘉
机构：京东科技信息技术有限公司
国家：中国
组别：产业组

AUTHOR : Xu Zhouping, Zhang Can, Huang Jia
UNIT : JD Technology Information Technology Co., Ltd.
COUNTRY : China
GROUP : Product Group

专家点评

京东智造云新品孵化平台这是一款基于精准消费需求和前沿人工智能技术
的一体化产品创新研发设计应用服务平台。它依托海量消费数据和智能分
析工具，在新品上市的各个环节精准满足个性化消费需求，实现柔性制造
下的千人千面。

EXPERT REVIEWS

The new product Incubation platform of JD zhizaoyun is an integrated product
innovation, R&D, design and application service platform based on precise
consumer demand and cutting-edge AI technology. Relying on massive consumption data and intelligent analysis tools, it accurately meets individualized
consumer demand in all aspects of new product launch and caters for the
expectations of all consumers under flexible manufacturing.

BUBBL

Bubbl 通过新的内容分发和社区参与工具颠覆了英国的零售和城镇中心管理部门。他们的软件虽然为应用程序添加了复杂的功能，但与应用程序分离，易于即插即用。这意味着沉浸式内容可以通过位置触发，将任何应用程序变成一个新的频道。

Bubbl disrupting the UK's retail and town centre management sectors with new content distribution and community engagement tools. Their software adds complex functionality to apps but is decoupled and easy to plug&play. This means immersive content can be triggered by location to turn any app into a new channel.

作者：Jo Eckersley, Mark Hunter, Taz Hossain
机构：Bubbl
国家：英国
组别：产业组

AUTHOR：Jo Eckersley, Mark Hunter, Taz Hossain
UNIT：Bubbl
COUNTRY：UK
GROUP：Product Group

专家点评

Bubbl 是一款基于移动设备，通过地理位置触发沉浸式内容的应用开发工具。它的出现定义了基于位置的移动营销新方式，让商家能更精准、更高效地铺设定制化内容。

EXPERT REVIEWS

Bubbl is an application development tool that is based on a mobile device and triggers immersive content through geographic location. Its emergence defines a new location-based mobile marketing method, allowing merchants to more accurately and efficiently lay out customized content.

科大讯飞智能助听器
IFLYTEK EARASST HA-01

这是一款以耳机为原型、以10%的价格实现了万元级助听器90%功能的数字通道助听器。当今全球共有听障患者3.6亿人，然而实际助听器普及率不到3%，大多数人无法享受到科技带来的便利。因此，市面上急需一款便宜有效、验配方便、跳脱出医疗设备式外观的产品。IFLYTEK EarAsst 正是为此而打造。

This is a digital channel hearing aid based on earphones, allowing consumers to realize 90% of the functions of a ¥10000 hearing aid at 10% of the price.Today, there are 360 million hearing-impaired patients in the world, but the actual penetration rate of hearing aids is less than 3%; most people cannot enjoy the convenience brought by technology.Therefore, the market urgently needs a product that is cheap, effective, easy to fit and does not have the appearance of a medical device. IFLYTEK EarAsst was developed in response to this.

作者：赵鸿 诸臣 许宝月
机构：科大讯飞股份有限公司
国家：中国
组别：产业组

AUTHOR : Zhao Hong, Zhu Chen, Xu Baoyue
UNIT : IFLYTEK Co., Ltd.
COUNTRY : China
GROUP : Product Group

专家点评

科大讯飞智能助听器以10%的价格实现了万元级助听器90%的功能，并实现了在家即可自行配置。通过引入时尚耳机的设计语言，在视觉上帮助患者融入大众人群，给予了听障人士更多的隐私和自在，享受到科技带来的便利。

EXPERT REVIEWS

IFLYTEK EarAsst HA-01 realizes 90% of the functions of a ¥ 10000 hearing aid at 10% of the price and is configurable by users at home. By introducing the design language of fashionable headphones, it visually helps patients integrate into the public, creates more privacy and freedom for hearing impaired people and lets them enjoy the convenience brought by science and technology.

RAPID RESCUE

"RAPID RESCUE"是一款简单美观、可挂墙收纳的担架。它被设计成放置在人群聚集的公共场所而不会让人感到不适，从而能够在紧急情况下做出快速反应。担架不是放置在储物箱中，而是将担架主体折叠后嵌入墙内，以达到紧凑的效果。主体重量轻，只需几个动作即可操作。其强度不亚于紧急情况下使用的担架。

"RAPID RESCUE"is a simple and beautiful stretcher that can be stored on a wall. It is designed to be placed in public places where people gather without discomfort, enabling rapid response in emergency situations. Instead of placing the stretcher in a storage box, the main body of the stretcher is folded and embedded in the wall to achieve compactness. The main body is lightweight and can be operated with a few actions. It is as strong as those used in emergency situations.

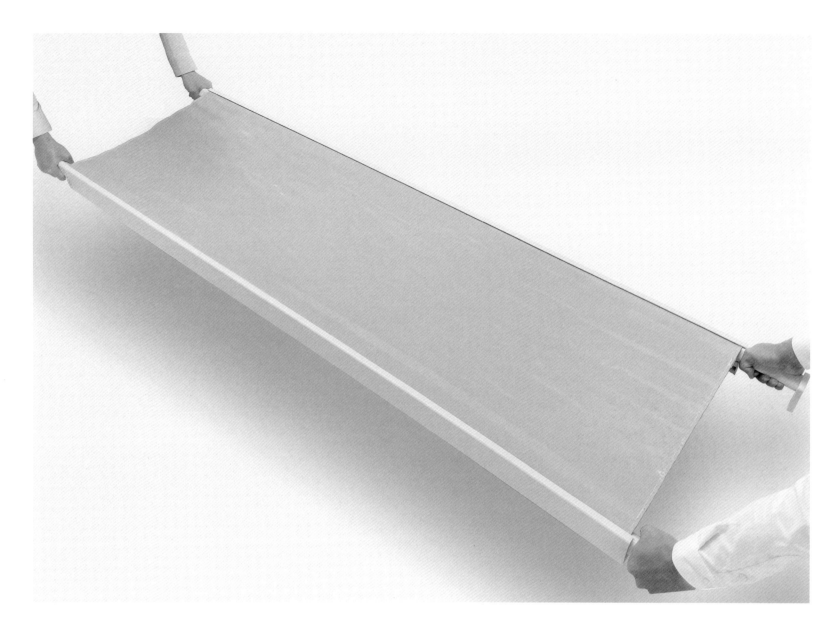

作者：Koji Syozaki, Motohiro Yanagishita, Haruka Nohara,
Toshikazu Tagawa
机构：UNION CORPORATION JAPAN, Kume Sekkei Co., Ltd.
国家：日本
组别：产业组

AUTHOR : Koji Syozaki, Motohiro Yanagishita, Haruka
Nohara, Toshikazu Tagawa
UNIT : UNION CORPORATION JAPAN, Kume Sekkei Co., Ltd.
COUNTRY : Japan
GROUP : Product Group

专家点评

Rapid Rescue 是为公共空间设计的应急救援担架。担架主体折叠并嵌入墙中，设计简洁美观兼具导视功能，便于在紧急情况下第一时间被发现并应用于救援。这是一款洋溢着生命关怀、未雨绸缪的设计产品。

EXPERT REVIEWS

Rapid Rescue is an emergency rescue stretcher designed for public space. The main body of the stretcher is folded and inlaid in the wall. The design is concise and beautiful and also has a guiding function to facilitate the earliest discovery and use in rescue in case of emergency. It is a design product that is permeated with caring for life and planning for a rainy day.

铜奖
2022 DIA
BRONZE
AWARD

丽江全景观光旅游列车
LIJIANG SIGHTSEEING TRAM

该车是全球首列全景观光山地旅游列车，首次采用全动力铰接转向架与大曲面电动调光玻璃，强适应环境，绿色环保，节能降耗。外观造型圆润饱满，打造专属梦幻蓝，纯净空灵。内饰设计融入东巴文化、纳西族精神内核和丽江美景，打造豪华、自然、民族三个不同美学概念车厢，为交通增值，实现文化＋交通以及文化＋旅游。

The train is the world's first panoramic mountain tourism train, the first use of full power articulated bogie and large curved electric dimming glass, strong adaptation to the environment, green environmental protection, energy saving and consumption reduction.The appearance is round and full, creating exclusive dream blue, pure and ethereal. The interior design integrates Dongba culture, Naxi spirit core and Lijiang beauty to create three different aesthetic concepts of luxury, natural and ethnic carriages, which add value to traffic and realize culture + traffic and culture + tourism.

作者：李鹏飞 张若曦 王湘翼 柳晓峰 钟飞
机构：中车株洲电力机车有限公司
国家：中国
组别：产业组

AUTHOR : Li Pengfei, Zhang Ruoxi, Wang Xiangyi, Liu Xiaofeng, Zhong Fei
UNIT : CRRC Zhuzhou Locomotive Co.,Ltd.
COUNTRY : China
GROUP : Product Group

专家点评
它是中国首列全景观光山地旅游列车，从轻奢、自然、民族三个不同美学概念，将传统交通工具设计转化为一趟"奇妙旅行"的体验设计。

EXPERT REVIEWS
It is China's first panoramic sightseeing mountain tourist train, which transforms the design of traditional transportation into an experience design of "a wonderful trip" from three different aesthetic concepts of entry luxury, nature and ethnicity.

U4电动自行车

U4是一款造型独特、通用性强、安全舒适、轻便易操作的高性价比电动自行车。它将更多功能和实用特性巧妙地集成在有限空间内，塑造出了纯粹的设计。专为城市通勤及休闲出行而打造。独特的车架几何设计，男女均可骑行。

U4 is a cost-effective electric bicycle with unique shape, strong versatility, safety and comfort, light and easy to operate. It cleverly integrates more functions and practical features into a limited space, creating a pure design. Built for urban commuting and leisure travel.Unique frame design works perfectly for both men and women.

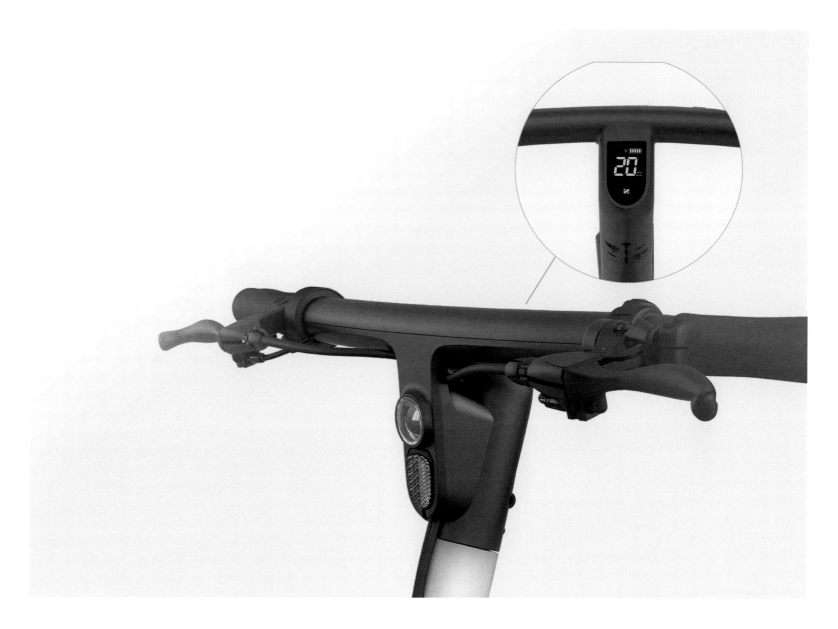

作者：刘丽 赵小伟 吴俊 刘兴举 陆其兵
机构：洪记两轮智能交通工具有限公司
国家：中国
组别：产业组

AUTHOR : Liu Li, Zhao Xiaowei, Wu Jun, Liu Xingju, Lu Qibing
UNIT : Hongji Intelligent Bike Co.,Ltd.
COUNTRY : China
GROUP : Product Group

专家点评

它实践了内走线的设计思路，让电源线、刹车线在电池管的周围穿隙而过。浑然一体的外观构造，彰显了大道至简、少即是多的美学。

EXPERT REVIEWS

It practices the design idea of internal wiring, so that the power line and brake line pass through the gap around the battery tube. The integral appearance structure highlights the aesthetics of the greatest truths being the simplest, and less being more.

移动看家数字乡村关怀版
ELDERLY ORIENTED VERSION OF MOBILE CARETAKER&DIGITAL VILLAGE

"移动看家数字乡村关怀版"是一款针对农村老年人的适老化安防产品系统。2022年农村老年人口（≥60岁）的比重达到了20.04%，这宣告我国农村已正式进入"老龄化社会"。2021年底中国移动集团正式启动"数字乡村与智慧社区大会战"专项行动，以安防业务为切入点全面融入国家和集团战略，"移动看家数字乡村关怀版"应势而生。

"Elderly Oriented Version of Mobile Caretaker&Digital Villag" is an age-friendly safety and security product system designed for elderly people in rural areas. This innovation comes in response to the demographic shift in China,where, by 2022, the elderly population (aged 60 and above) in rural regions had reached a significant 20.04%, signifying the country's transition into an "aging society" in rural areas. Notably, in late 2021, China Mobile Group initiated the "Digital Village and Smart Community Campaign," to incorporate security service into both the state's and the corporate's strategy. "The Elderly Oriented Version of Mobile Caretaker&Digital Villag " was developed in response to these trends.

作者：王嘉榕 杨易 靳倩倩 程宝平 王欣
机构：中国移动杭州研发中心
国家：中国
组别：产业组

AUTHOR : Wang Jiarong, Yang Yi, Jin Qianqian, Cheng Baoping, Wang Xin
UNIT : China Mobile(Hangzhou) Information Technology Co., Ltd.
COUNTRY : China
GROUP : Product Group

专家点评

它在硬件和软件的双重支持下，基于广泛的样本调研和大数据分析，运用 AI 监测、视频监控等联防联控手段，完成不同村、镇、县政府的定制化服务。关怀今天的老人，就是关怀明天的自己。

EXPERT REVIEWS

With the dual support of hardware and software, and based on extensive sample research and big data analysis, it employs AI monitoring, video surveillance and other joint prevention and control means to complete customized services for different village, town and county governments. To care for the elderly today is to care for yourself tomorrow.

问界 M5 EV
AITO M5 EV

AITO 问界 M5 EV 是 AITO 首款智慧豪华纯电 SUV，整体设计延续 AITO 品牌极致、简约、纯净的理念，打造时尚前卫、经典耐看的都市生态美学。同时，秉持以人为本的设计初心，内饰贴心的细节设计、宽敞舒适的座舱空间，为消费者提供高品质舒适座舱。搭载 HUAWEI DriveONE 纯电驱平台，性能与续航兼备，搭载最新的 HarmonyOS 3.0 系统，从 HiCar 到 HarmonyOS 深度融合，打造卓越驾乘体验。

The AITO M5 EV is AITO's first smart luxury pure electric SUV. The design concept is ultimate, simple and pure, which is to create fashion-forward, classic and urban ecological aesthetics. What provides consumers with high-quality and comfortable cabins is people-oriented design, the interior intimate details design and spacious and comfortable cabin design. The HUAWEI DriveONE platform provides both performance and battery life. With the latest HarmonyOS 3.0 system, HiCar and HarmonyOS are deeply integrated to create a superior driving experience.

作者：伍国平 聂晶晶 谭燕婷 冉思杰 姚国栋
机构：华为技术有限公司 / 重庆金康赛力斯有限公司
国家：中国
组别：产业组

专家点评

它能够完成超级桌面、手表互联的全场景协同工作，同时让车机成为手机生态服务的拓展平台，实现手机 APP 的无缝流转。它是设计与功能通力合作、机能与美学默契呼应的完美标杆。

AUTHOR : Wu Guoping, Nie Jingjing, Tan Yanting, Ran Sijie, Yao Guodong
UNIT : Huawei Technologies Co., Ltd. / Chongqing Jinkang Celis Co., Ltd.
COUNTRY : China
GROUP : Product Group

EXPERT REVIEWS

It can complete full-scenario collaboration of super desktop and watch inter-connection and enables in-vehicle devices to become an expansion platform for mobile phone ecosystem services, bringing about seamless interaction flow of mobile apps. It is a perfect benchmark for the cooperation between design and function, and the tacit response between function and aesthetics.

COFFEE KREIS

Kreis 杯是一款可持续、可重复使用、可生物降解的咖啡杯，由回收的咖啡渣制成。我们提供一次性塑料／纸杯的替代品，并减少每年被扔到垃圾填埋场的数百万吨咖啡渣。我们的产品采用双层隔热材料，在保温或保冷方面具有极大的优势。

The Kreis Cup is a sustainable, reusable and biodegradable coffee cup made of recycled coffee grounds. We provide an alternative to single use plastic/paper cups and to reduce millions of tons of coffee waste that are thrown to the landfills every year. Our products are double walled insulated, which provides great advantages to retain heat or cold.

作者：Ricardo Garcia
机构：Coffee Kreis
国家：美国
组别：产业组

AUTHOR : Ricardo Garcia
UNIT : Coffee Kreis
COUNTRY : United States
GROUP : Product Group

专家点评

它将留待回收的咖啡渣，创造成一个崭新的咖啡杯。设计师在赋予咖啡二次生命的同时，也就此展开了变废为宝的奇妙造物之旅。

EXPERT REVIEWS

It makes a new coffee cup using the coffee grounds to be recycled. While giving coffee a second life, the designer also unfolds a wonderful journey of turning waste into treasure.

兰亭续
LANTING XU

《兰亭续》实现跨行业首创，将文化、艺术、科技相结合．将《兰亭序》织成古锦，采用现代织锦五色交织法，优化传统工艺，实现可复制性。与独创的"冰岛水晶"相融合，塑造在河水中被千古冰封的样子。实现书法美、织锦美、意境美和谐统一。延续《兰亭序》以文会友的生活方式，倡导当下以茶会友的健康生活方式，让艺术走进生活。

"Lanting Xu" is a cross-industry innovation that combines culture, art, science and technology."Lanting Xu" (Preface to the Poems Composed at the Orchid Pavilion), a famous Chinese calligraphy work is woven into ancient brocade, using the modern brocade five colors interweaving method, optimizing the traditional process, to achieve replicability. It incorporates the unique "Icelandic crystal" to create the appearance of being frozen in ancient river waters. This achievement harmoniously unites calligraphy, brocade weaving, and artistic conception, continuing the life style of social gatherings centered around art as described in "Lanting Xu" . It encourages a healthy lifestyle by suggesting modern tea gatherings as a way to bring art into everyday life.

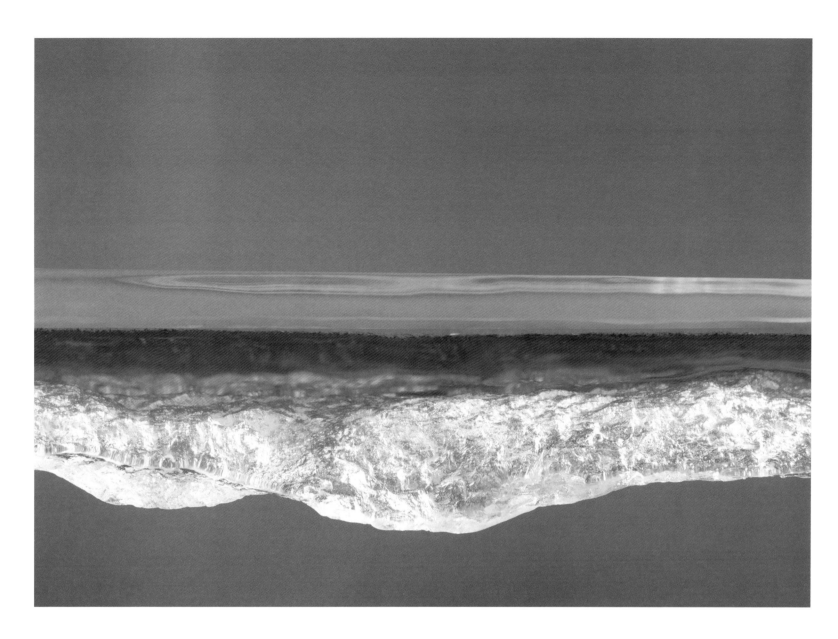

作者：李加林 陈刘成 白雅静
机构：浙江茶荟家居有限公司
国家：中国
组别：产业组

AUTHOR : Li Jialin, Chen Liucheng, Bai Yajing
UNIT : Zhejiang ChahuiJiaju Company
COUNTRY : China
GROUP : Product Group

专家点评

它一改高端茶桌采用名贵木材的思维定式，突破尺寸限制、高温卷曲、色彩变化等诸多难题，把现代织锦五色交织工艺重塑的《兰亭序》，封装于"冰岛水晶"的桌面内部。它是对日常生活审美化的又一次成功践行。

EXPERT REVIEWS

It changes the mindset of high-end tea tables using precious wood, breaks through many problems of size limitation, high temperature curling, color change, etc., and encapsulates the "Orchid Pavilion" remolded by modern brocade five-color interweaving technology inside the desktop of "Icelandic Crystal". It is another successful practice of beautification of daily life.

S 计划 MODEL S
SPLAN MODEL S

这是一款采用创新电子配重技术的家用智能力训器，通过内置前沿的 AI 算法，确保用户全程使用安全。结合强大的伺服电机、智能系统、钢化玻璃全身镜及 S Plan 的5S（Shape Plan、Slim Plan、Strong Plan、Stretch Plan、Simple Plan）品牌诉求，为用户带来更安全、更有效，同时也更易于长期坚持的系统性身材管理方案。

This is an intelligent training device for home using innovative electronic counterweight technology, with the built-in cutting-edge AI algorithm to ensure the safety of users throughout the process. Combining powerful servo motors, intelligent systems, tempered glass full-length mirrors and S Plan's 5S (Shape Plan, Slim Plan, Strong Plan, Stretch Plan, Simple Plan) brand appeals, bringing users a safer, more effective systematic body management plan that is easy to adhere to for a long time.

作者：关勇超
机构：北京觅淘智联科技有限公司
国家：中国
组别：产业组

AUTHOR : Guan Yongchao
UNIT : Beijing Xburn technology Co.,Ltd.
COUNTRY : China
GROUP : Product Group

专家点评

它将传统的物理配重直接迭代到电子配重时代，打破了健身房的环境和专业壁垒，实现了健身场景的随时切换和专业教练的远程在场。

EXPERT REVIEWS

It directly iterates the traditional physical counterweight to the era of electronic counterweight. It breaks the environmental and professional barriers of the gym, and realizes the switching of fitness scenes at any time and the remote presence of professional coaches.

一棵
ECHO

"一棵"通过把所有大棚环境参数（光照、水分温湿度、土壤温湿度、EC 指数）进行综合数据算法建模，将大棚建设、管护、调整、流转、估价等各个环节通过数据进行科学管理，实现 1 人可管多棚，光照灌溉等环境调整通过手机小程序即可进行科学判断，摆脱单纯依靠传统经验，提升科学管理。种下"一棵"，降低大棚人工成本，科学提升作物产量。

"Echo" conducts comprehensive modeling using environmental data and gives advice such as greenhouse construction, management and maintenance. Therefore, one farmer can manage multiple greenhouses, and can make scientific judgments through the mobile App. "Echo" helps reduce the labor cost of the greenhouse, and scientifically increases the crop productivity.

作者:张昊 黎欣 王琴琴 王晓辉 张超峰
机构:东方智感（浙江）科技股份有限公司
国家:中国
组别:产业组

AUTHOR : Zhang Hao, Li Xin, Wang Qinqin, Wang Xiaohui, Zhang Chaofeng
UNIT : Insentek
COUNTRY : China
GROUP : Product Group

专家点评

它能和作物一同感知重要的生存元素，并通过算法归纳生成一份清晰简洁的大棚能效表。读懂对方的最好方式，是成为对方，用"一棵"，理解每一棵作物。

EXPERT REVIEWS

It can perceive the important survival elements together with the crop, and generate a clear and concise table of greenhouse energy efficiency through algorithm induction. The best way to know the other party is to become the other party. "ECHO" is used to understand every crop.

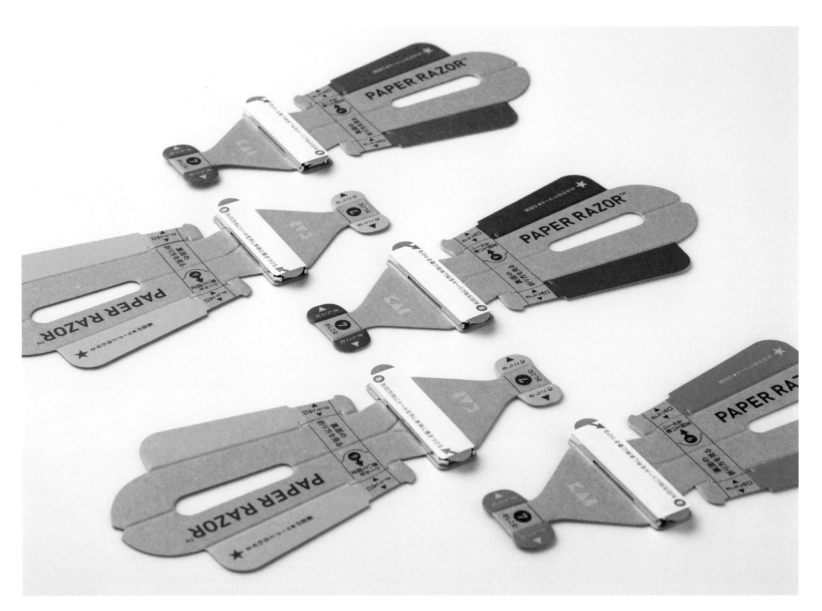

PAPER RAZOR

一款采用纸柄和金属头设计的剃须刀，以减少塑料的使用。这款剃须刀采用扁平包装，组装后即可使用，厚度仅为3毫米，重量仅为4克。印刷图案可以改变，从而创造出各种图案。这是一个全新的概念，产品除了剃须的基本功能外，还提供了享受选择和享受使用的情感价值。

A shaver designed with a paper handle and metal head to reduce plastic use. It is packed flat and is ready for use after assembly, with a thickness of just 3 millimetre and a weight of just 4 gram. The printed graphic can be changed to create a variety of patterns. This is a new concept that provides the emotional value of the enjoyment of selection and the enjoyment of use in addition to the basic function of shaving.

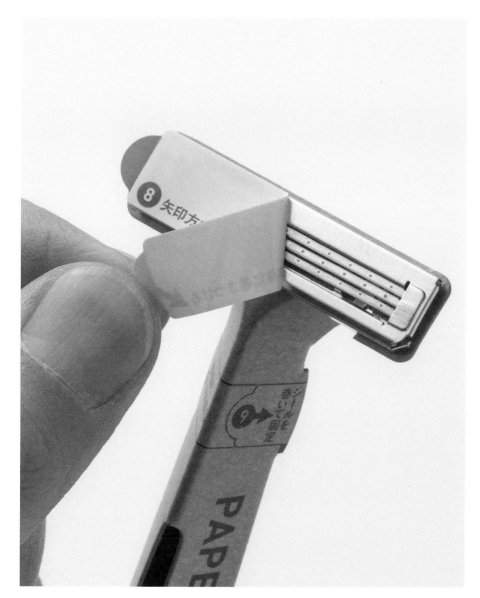

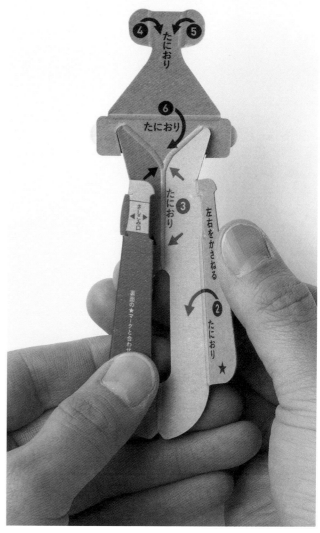

作者：Shunsuke Shioya, Mai Kadokura
机构：Kai Industries Co.,Ltd. Kai Corporation
国家：日本
组别：产业组

AUTHOR : Shunsuke Shioya, Mai Kadokura
UNIT : Kai Industries Co.,Ltd. Kai Corporation
COUNTRY : Japan
GROUP : Product Group

专家点评

它运用稳固的折纸结构把柔软的纸和锋利的金属巧妙融合，实现了可持续性和便携性的完美结合。

EXPERT REVIEWS

It uses a steady origami structure to cleverly integrate soft paper and sharp metal, achieving a perfect combination of sustainability and portability.

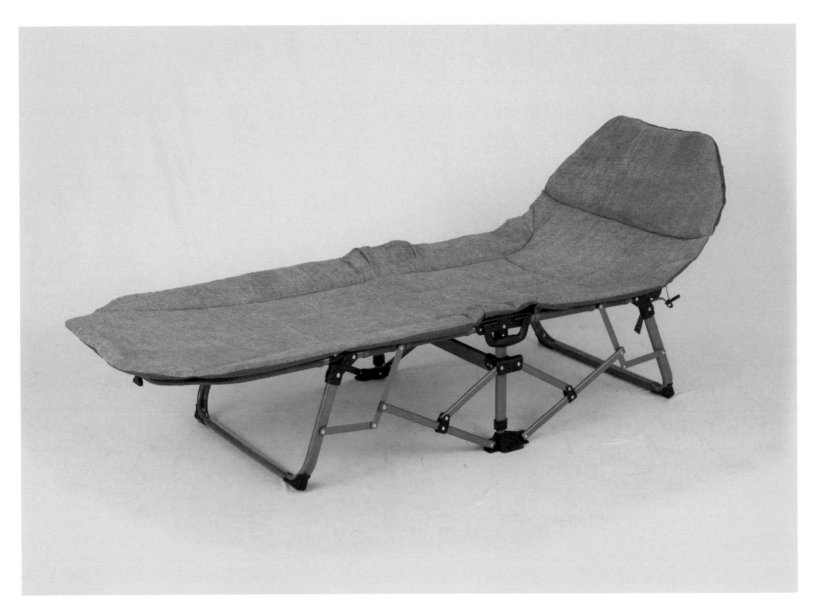

DIA
DESIGN
INTELLIGENCE
AWARD

BRONZE
AWARD
铜奖

快速折叠床
SPEEDY COT

这是快速收折机构，收折后体较小；靠背可五档调节，坐躺两用，满足不同的使用需求。框架采用了镁合金金属材料，重量轻、强度好、环保；面料是柔软舒适的双彩绒，床面的造型也柔美大方，采用云朵造型作为曲线设计元素，对软垫进行了巧妙的裁剪分割，可给用户带来一种视觉上的舒适；同时软垫内填充高密度的回弹海绵，饱满柔软。

Speedy Cot features a quick folding mechanism that reduces its size when folded. The backrest can be adjusted to five different positions, allowing it to serve as both a seat and a bed to meet various usage needs. The frame is constructed from strong, lightweight and environmentally friendly magnesium alloy. The fabric is made from soft and comfortable dual-color velvet. The design of the bed surface is elegant, incorporating cloud-shaped elements into its curved design. The cushioning has been cleverly cut and segmented to provide users with visual comfort. Additionally, the cushion is filled with high-density rebound sponge for full softness.

作者：余红飞 卢小明
机构：浙江泰普森实业集团有限公司
国家：中国
组别：产业组

AUTHOR : Yu Hongfei, Lu Xiaoming
UNIT :Zhejiang Hengfeng Top Leisure Co.,Ltd.
COUNTRY : China
GROUP : Product Group

专家点评

它有着轻便、强度的镁合金材质和高密度的回弹海绵，两秒即可完成开合。在快节奏的都市生活中，它让每一次弥足珍贵的小憩都轻松适意。

EXPERT REVIEWS

It is made of light and strong magnesium alloy and high-density resilient sponge and can be folded and opened in two seconds. In the fast-paced urban life, it makes every precious nap easy and comfortable.

明日之星
2022 DIA
FUTURE TALENTS
AWARD

PROJECT MOAB

Moab 项目是一个硬件套件，旨在通过将有形的人工智能体验带到用户的桌子上，让用户感受到机器教学的强大功能。Moab 是一款球平衡机器人，它利用计算机视觉和各种控制方法来平衡物体。用户可以选择 PID 循环等经典控制系统、训练有素的人工智能控制系统，或通过集成操纵杆进行手动控制。机器人的反应和"技能"取决于控制方法，用户可以根据自己训练的模型进行测试和评估。硬件套件与基于网络的机器教学服务相结合，用户可以创建并训练自己的人工智能模型或"大脑"。

Project Moab is a hardware kit designed to onboard users to the power of machine teaching by bringing a tangible AI experience right to their desk. Moab, a ball-balancing robot, leverages computer vision and various control methods to balance objects. Users can select classical control systems such as PID loops, trained AI control systems, or manual control via an integrated joystick. The robot's reactions and "skill" depends on the control method, which users can test and evaluate against their own trained models. The hardware kit pairs with a web-based machine teaching service, where users create and train their own AI models or "brains".

作者：Scotty Paton, Nissa Van Meter, Avinash Singh, Kyle Skelton, Scott Stanfield
机构：Microsoft (BusinessAI) / FreshConsulting
国家：美国
组别：概念组

AUTHOR : Scotty Paton, Nissa Van Meter, Avinash Singh, Kyle Skelton, Scott Stanfield
UNIT : Microsoft (BusinessAI) / FreshConsulting
COUNTRY : United States
GROUP : Concept Group

专家点评

它是一个极简的硬件套件，却能够孕育出控制、超越物体平衡的思维。它让工程师在一个有关需求与限制的智能控制系统中，"戴着镣铐舞蹈"，不断测试和评估自己的训练模型。

EXPERT REVIEWS

It is a minimalist hardware kit that can foster thinking about controlling and transcending object balance. It lets engineers "dance in shackles" in an intelligent control system about demands and limitations to constantly test and evaluate their training models.

低碳校园
LOW-CARBON CAMPUS

低碳校园由行为设计驱动,将数据与场景智慧协同,通过艺术生产创造低碳新经济,推动低碳成为社会新风尚。提出"学院即社区,低碳即时尚"的设计主张,强调低碳校园建设不应只关注校园能耗的减碳排量,更应通过行为设计驱动校园师生主动践行低碳行为,让每一个人都成为低碳理念的践行者与传播者。

Low-carbon campus is driven by behavioral design. The project aims at promoting low-carbon as a new social trend by making art production a new low-carbon economy and using the combination of data and scenarios. It advocates the idea that "the campus is a community, and being low-carbon is in vogue," emphasizing that the construction of a low-carbon campus should not only focus on reducing carbon emissions from campus energy consumption but should also actively encourage students and staff to adopt low-carbon behaviors through behavioral design. This way, everyone becomes a practitioner and promoter of the low-carbon concept.

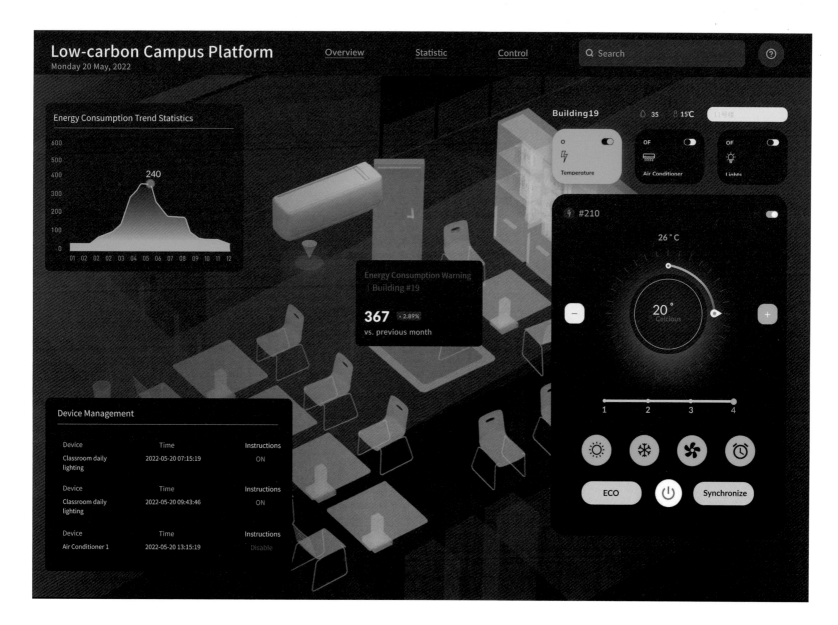

作者：陈赟佳 张璐璐 冯瑞云 丁雨欣 蔡和睿
机构：中国美术学院 / 阿里云计算有限公司
国家：中国
组别：概念组

AUTHOR : Chen Yunjia, Zhang Lulu, Feng Ruiyun, Ding Yuxin, Cai Herui
UNIT : China Academy of Art / Alibaba Cloud Computing Co.,Ltd.
COUNTRY : China
GROUP : Concept Group

专家点评

它依托物联采集，将校园日常场景下的低碳行为，转化成碳排放数据，同步上传校园能耗数据管理平台。它让碳中和的实践持续在社群里发酵，成为校园新风尚。

EXPERT REVIEWS

It relies on IoT acquisition, converts the low-carbon behaviors on daily scenes of the campus into carbon emission data and synchronously uploads it to the campus energy consumption data management platform. It allows the practice of carbon neutrality to ferment continuously in the community and become a new trend on campus.

设计新锐
2022 DIA
YOUNG TALENTS
AWARD

GRAVEL - A FOOTBALL SHOE FROM GHANA

Gravel 是一款足球鞋，专为加纳足球场的恶劣条件而设计。这款足球鞋由大麻鞋面和可更换的橡胶外底组成。其创新的鞋带系统将可更换的橡胶外底与大麻织物鞋面巧妙连接起来。这个项目的基础是一次穿越加纳的研究之旅。

Gravel is a football shoe, which was designed for the rough conditions of Ghanaian football pitches. It consists of a hemp upper and a replaceable rubber outsole. An innovative lacing-system connects the replaceable rubber outsole to the hemp-textile upper. The foundation of this project is a research trip through Ghana.

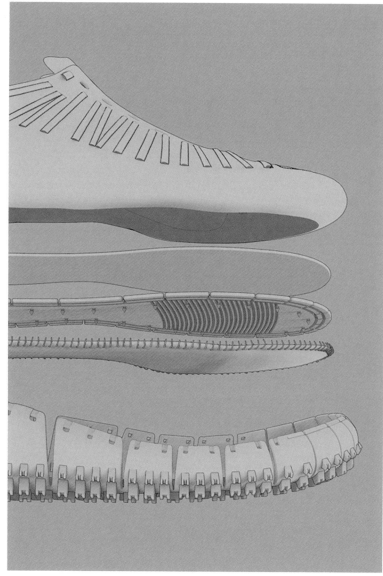

作者：Ruben Asuo
国家：奥地利
组别：概念组

AUTHOR : Ruben Asuo
COUNTRY : Austria
GROUP : Concept Group

专家点评

它由大麻鞋面和可更换的橡胶外底组成，其创新的鞋带系统将可更换的橡胶外底与大麻织物鞋面巧妙连接，刚性兼具弹性的内底，带有小鞋钉，以应对球场上的苛刻动作。即便是沙土和岩石的地面，穿上它也能快乐地享受足球运动。

EXPERT REVIEWS

It consists of a hemp upper and a replaceable rubber outsole. An innovative lacing-system connects the replaceable rubber outsole to the hemp-textile upper. A rigid but still flexible insole with little spikes coped with the demanding actions on the field. Even if the ground is sandy and rocky, we can still enjoy the football game when wearing it.

碳交易体验平台动物园站 —— COZoo
CARBON TRADING EXPERIENCE PLATFORM — COZOO

"COZoo" 动物园站是一个碳交易体验平台，致力于提高大众对碳交易的认知，并构建绿色经济的生活方式。COZoo 围绕 "生态孤岛 + 你 = 生态共同体" 的核心理念展开，意在通过碳交易体验设计、艺术赋能和品牌升级三个方面推动减污降碳协同增效，促进经济社会发展的绿色转型。

"COZoo" is a carbon trading experience platform that aims to raise public awareness of carbon trading and build a green economy lifestyle. Through carbon trading experience design, art empowerment and brand upgrading, CO-Zoo aims to promote synergy and efficiency in reducing pollution and carbon emissions, and to facilitate the green transformation of economic and social development.

作者：景斯阳 倪尔璐 黎超群 王晓彤 王琪
机构：中央美术学院设计学院
国家：中国
组别：概念组

AUTHOR : Jing Siyang, Ni Erlu, Li Chaoqun, Wang Xiaotong,
Wang Qi
UNIT : School of Design, Central Academy of Fine Arts
COUNTRY : China
GROUP : Concept Group

专家点评

它致力于提高大众对碳交易的认知，并构建绿色经济的生活方式。它旨在通过碳交易体验设计、艺术赋能和品牌升级三个方面推动减污降碳协同增效，促进经济社会发展的绿色转型。

EXPERT REVIEWS

It is dedicated to raising public awareness of carbon trading and building a green economic lifestyle. It aims to promote the green transformation of economic and social development through carbon trading experience design, artistic empowerment and brand upgrading, to promote synergy in reducing pollution and carbon emission.

RHAETUS —— 三轮折叠货物载具
RHAETUS — ELECTRIC FOLDING CARGO TRIKE

市面上的 cargo bike 前方有足够的置物区可安心承载货物、宠物，甚至是小孩，但却因为车身庞大，在载物以外时不方便使用。 Rhaetus 能收折前方置物区的设计便解决了此问题，在载物以外的使用上能将车身转变为轻型载具，增加载具的移动性、机动性及使用性。

The cargo bike in the market has storage space in the front to carry cargo, pets, and even children, but it is not suitable for general use beyond cargo carrying because of its large size. Rhaetus' folding front storage area design solves this problem by transforming the bike into a light carrier for use beyond a cargo bike, increasing the bike's mobility, functionality, and usability.

作者：韩皓宇 李承浚
国家：中国（台湾）
组别：概念组

AUTHOR : Han Haoyu, Li Chengjun
COUNTRY : Taiwan, China
GROUP : Concept Group

专家点评

它除了能够满足装载货物的基本要求，置物区的可收折处理，使得它可以轻松转变为轻型载具。设计的最终目的并非产品，而是为了满足人的需求。

EXPERT REVIEWS

In addition to meeting the basic requirements of loading cargo, the foldable design of the storage area turns it into a light carrier. The ultimate purpose of design is not to produce products, but to meet people's needs.

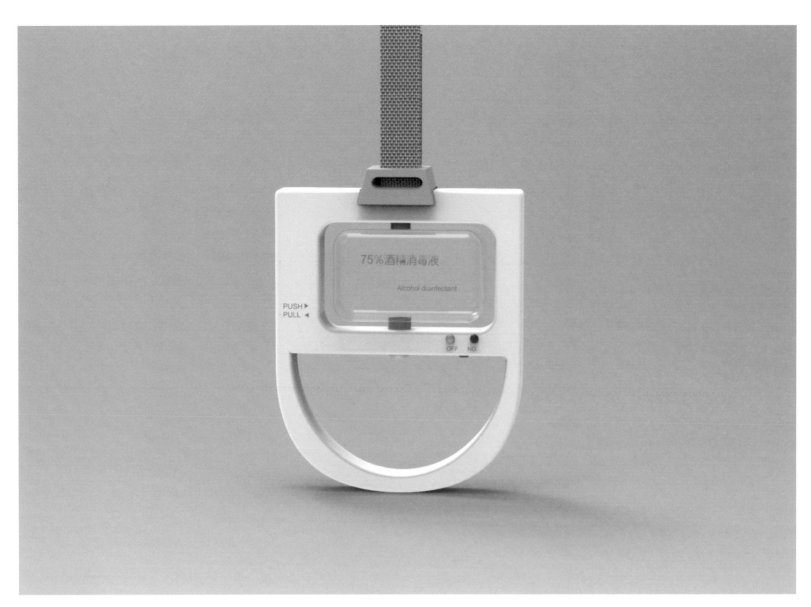

公共交通设施自动消毒拉手设计
AUTOMATICALLY DISINFECT HANDRAILS

公共交通设施自动消毒拉手是一款通过红外线感应对拉手进行自动消毒，解决了公共交通设计消毒问题的产品。利用红外线感应技术对拉手进行消毒，解决了公共交通设施的拉手消毒问题，抽屉式电源使得更换更为便捷。

The AUTOMATICALLY DISINFECT HANDRAILS is a product that uses infrared sensing to automatically disinfect handrails, addressing the issue of disinfection in public transportation design. Ensuring the disinfection of handrails in public transport facilities is crucial. The use of infrared sensor technology enables the automatic disinfection of handrails, and the drawer-type power supply makes replacement convenient and straightforward.

作者：陆凡 程禹 朱秀荣 吴蔓芊
国家：中国
组别：概念组

AUTHOR : Lu Fan, Cheng Yu, Zhu Xiurong, Wu Manqian
COUNTRY : China
GROUP : Concept Group

专家点评

它利用红外线感应技术，可以在人员手部接触后，迅速感知，并立即进行消毒。科技的进步，往往体现在细节上的温度。

EXPERT REVIEWS

It utilizes infrared sensing technology to quickly detect and disinfect upon contact with people's hands. Technological advancements often manifest in the warmth of details.

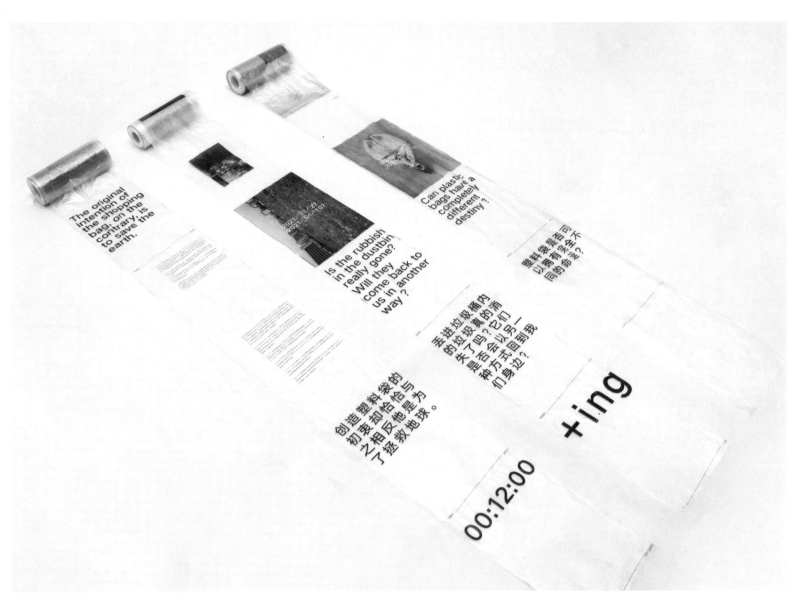

十二分钟
TWELVE MINUTES

以"人们使用塑料袋的平均时间为十二分钟"为出发点，聚焦人类对塑料袋的错误使用带来的环境问题。作品以摄影集的方式呈现，将塑料袋之书分为生产使用、丢弃污染、二次利用三个部分，让读者切实感受塑料袋从生产、使用再到丢弃的全过程。

Taking "the average time people use plastic bags is 12 minutes" as the starting point, we will focus on the environmental problems caused by human misuse of plastic bags.The work is presented in the form of a photography collection, dividing "The Book of Plastic Bags" into three parts: production and use, disposal and pollution, and secondary utilization. This allows readers to truly experience the entire process of plastic bags from production, usage, and disposal.

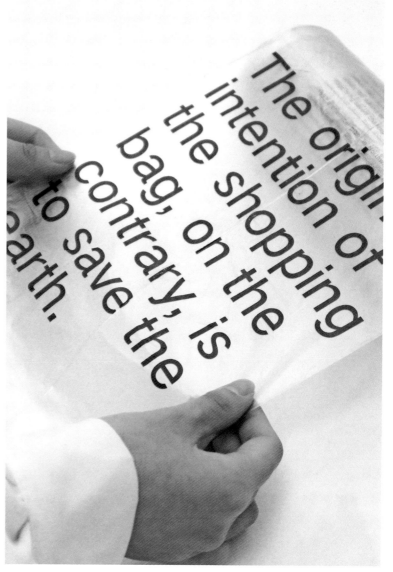

作者： 庞烨 向冰姿 彭玲 梁国健
国家： 中国
组别： 概念组

AUTHOR : Pang Ye, Xiang Bingzi, Peng Ling, Liang Guojian
COUNTRY : China
GROUP : Concept Group

专家点评

它引导观众反思，塑料污染的原罪，是塑料袋本身，还是人类的误用。科技进步带来便捷的同时，却诱导人类走向了污染的泥淖。

EXPERT REVIEWS

It leads the audience to reflect on the original sin of plastic pollution. Is it the plastic bag itself, or is it human misuse? While scientific and technological progress brings convenience, it also induces human beings to the mire of pollution.

可持续防汛净水沙袋
SANDBAGS FOR FLOOD PROTECTION

可持续净水防汛沙袋通过内部高密度海绵材质，将洪水快速吸入沙袋内部，同时将污染的洪水通过沙袋下方的污水净化器进行污水净化处理。最后将污染的洪水转换成干净卫生的饮水资源。本次设计有效解决传统沙袋吸水慢、排水效率低、搬运困难等问题，为受灾人群提供卫生的饮水资源，提高灾后生存环境条件。

The sustainable water purification and flood control sandbag uses a high-density sponge material inside to quickly inhale the floodwater in, and at the same time, the polluted floodwater will be purified by the sewage purifier under the sandbag. Finally, the polluted flood water will be converted into a clean and hygienic drinking water source. This design effectively solves the problems of slow water absorption, low drainage efficiency, and difficult transportation of traditional sandbags, provides hygienic drinking water resources for the disaster-stricken people, improves the living environment conditions of post-disaster areas, and ensures the physical and psychological health of the disaster-affected people.

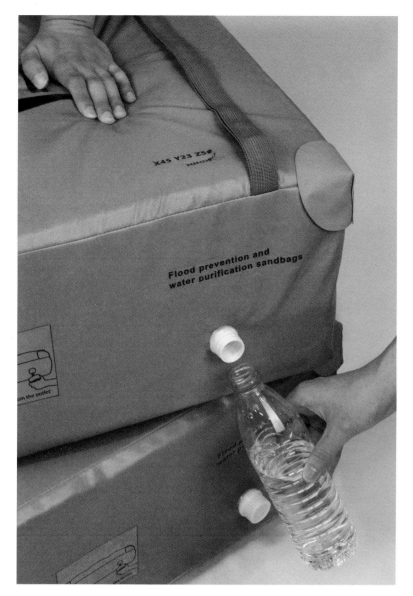

作者：田皓瑞 胡健祯 张驰
国家：中国
组别：概念组

AUTHOR : Tian Haorui, Hu Jianzhen, Zhang Chi
COUNTRY : China
GROUP : Concept Group

专家点评

它有效解决传统沙袋吸水慢、排水效率低、搬运困难等问题，为受灾人群提供卫生的饮水资源，提高灾后生存环境条件，保障了受灾人员的灾后生理和心理的健康。面对灾难，科技应该成为普通人最坚实的臂膀。

EXPERT REVIEWS

It effectively solves the problems of traditional sandbags such as slow water absorption, low drainage efficiency and difficult transportation, provides wholesome drinking water resources for the affected people, improves the living environment conditions of post-disaster areas, and ensures the physical and psychological health of the disaster-affected people. In the face of disaster, technology should become the most solid arm for ordinary people.

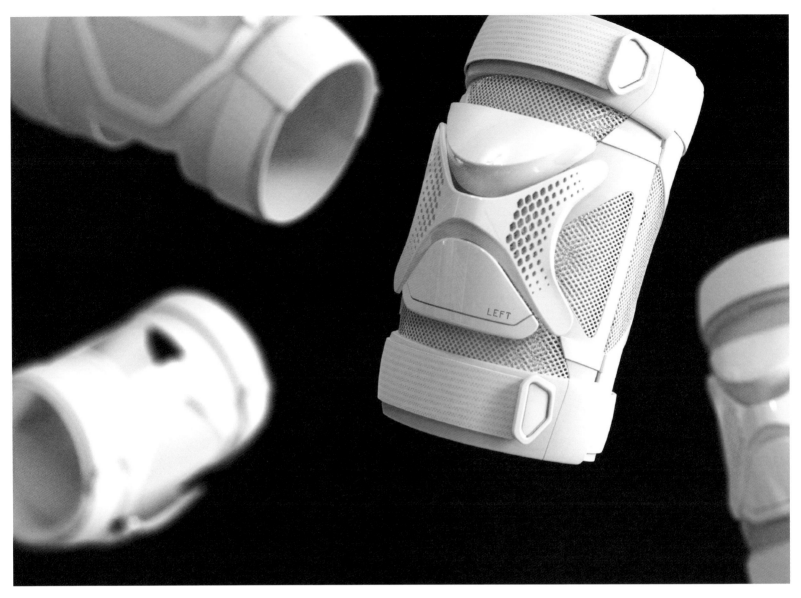

HALCÓN ——复合关节极限运动护具
HALCÓN — MULTI-JOINT EXTREME SPORTS BRACE

Halcón 是一组专为滑板玩家设计的护具，不管是硬壳还是布料的部分都有别于市面上的所有护具，直接针对传统护具的痛点来进行设计，目的是要创造集"灵活性""透气性""外观颜值"为一体的全新护具。Halcón 最大的创新点，是改变硬壳的连接方式，设计出一种三段式的关节，如此护具便会随着我们膝盖的活动而自动调节到合适的位置。

Halcón is a set of extreme sports protective gear that catered to young skateboarders. This is different from all the protective gears on the market nowadays, we design directly to address the pain points and aim to complete a new protective gear that integrates "flexibility", "breathability" and "aesthetic value".Moreover, Halcón's most important innovation is the improvement of the connection between the hard shell and the fabric. We designed a three-stage joint so that the protector automatically adjusts to the right position with the movement of our knee.

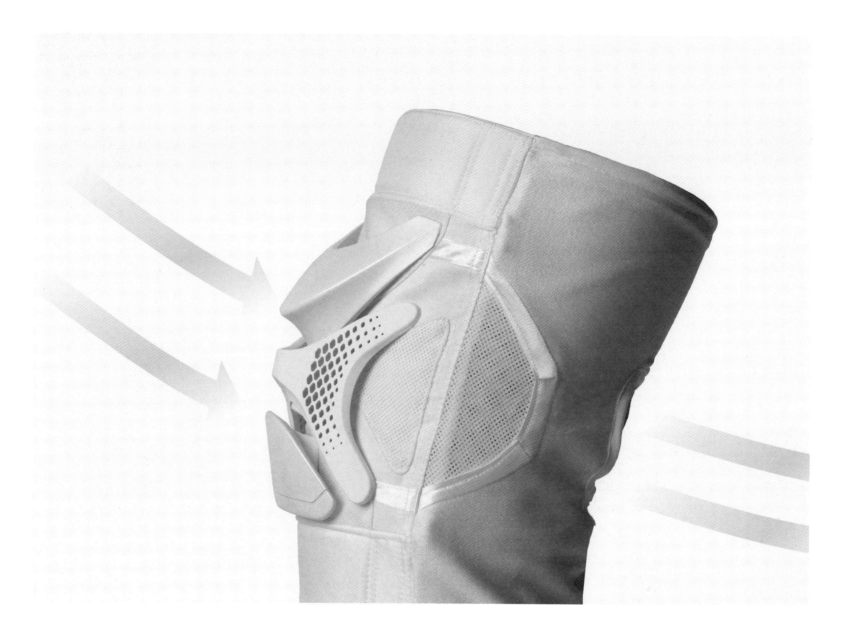

作者：陈冠桦
国家：中国（台湾）
组别：概念组

AUTHOR : Chen Guanhua
COUNTRY : Taiwan, China
GROUP : Concept Group

专家点评

它将耐磨、透气和穿戴灵活这三个特性集合于一身，从原本直接缝死在布料上的硬壳，改为三段式的固定，让膝盖关节在极限运动中自由活动。在科技的加持下，护具变成"皮肤"，不仅保护身体，也成为身体的延伸。

EXPERT REVIEWS

It combines the three features of durability, breathability, and flexibility, transitioning from the previous rigid shell directly sewn onto the fabric to a three-segmented fixation, allowing the knee joint to move freely during extreme activities. With the support of technology, this protective gear transforms into a "second skin", not only safeguarding the body but also becoming an extension of it.

佳作奖
2022 DIA HONORABLE MENTION AWARD

文化创新
CULTURAL INNOVATION

S3 漫步杯
S3 WALK CUP

CS CHAIR

作者：陈文 刘培清
机构：杭州膳佳家居用品有限公司
国家：中国
组别：产业组

AUTHOR : Chen Wen, Liu Peiqing
UNIT : Hangzhou Shanjia Lifestyles Co.,Ltd.
COUNTRY : China
GROUP : Product Group

作者：Yohei Yamamoto, Yasuhiro Nakamura, Yuki Takeya , Ryogo Morita, Kohei Wada
机构：ITOKI CORPORATION
国家：日本
组别：产业组

AUTHOR : Yohei Yamamoto, Yasuhiro Nakamura, Yuki Takeya , Ryogo Morita, Kohei Wada
UNIT : ITOKI CORPORATION
COUNTRY : Japan
GROUP : Product Group

S3通过原创专利滑块结构，解决了随手杯可单手轻松开启直饮口的问题，且单手开合可做到密封不漏水。

The S3 is an original patented slider structure, which solves the problem that the cup can be easily opened with one hand, and the opening and closing with one hand can be sealed and watertight.

CS 椅是一款优雅的休闲扶手椅，隐藏着诸多优点：令人惊奇的可折叠、可嵌套和可收纳，并且所有这些功能均可以单手操作。

CS chair is an elegant occasional armchair with hidden talents: surprisingly foldable, nestable, and storable, all with one hand.

GHOST CLOCK

ECHO 智能台灯
ECHO SMART DESK LAMP

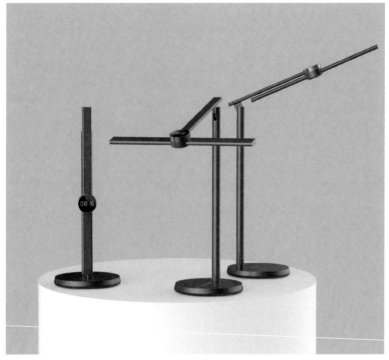

作者 : Wonyoung Kim, Minkyoung Bae, Hojin Lee
机构 : studio when
国家 : 韩国
组别 : 产业组

AUTHOR :Wonyoung Kim, Minkyoung Bae, Hojin Lee
UNIT : studio when
COUNTRY : South Korea
GROUP : Product Group

作者 : 刘吴旭
机构 : 奕至家居科技 (深圳) 有限公司
国家 : 中国
组别 : 产业组

AUTHOR : Liu Wuxu
UNIT : Shenzhen EZVALO Technology Company Limited
COUNTRY : China
GROUP : Product Group

"幽灵钟"是一款现代时钟, 根据空间和视线描绘出挂钟和座钟的灵活特性。 "幽灵钟"的设计理念是在称为时间的无形元素上覆盖一块布。

"Ghost clock"is a contemporary timepiece that depicts the supple identity of a wall clock and a table clock according to the space and gaze."Ghost clock"is designed from the concept of covering a cloth over the intangible element called time.

Echo 是一台通过多形态变化, 满足多场景阅读照明与氛围照明的智能台灯。

Echo is a smart desk lamp that meets the needs of multi-scene reading lighting and ambient lighting through multi form changes.

普洛
PRO

作者：康士坦丁·葛切奇
机构：上海阿旺特家具有限公司
国家：中国
组别：产业组

AUTHOR : Konstantin Grcic
UNIT : Shanghai Avarte Furniture Co.,Ltd.
COUNTRY : China
GROUP : Product Group

Pro 是通过座椅形态设计从而解决现代教学空间需求的一款学生椅。

Pro is a student chair that solves the needs of modern teaching space through the design of chair shape.

邬先生的灯
MR WU'S LAMP

作者：赵智峰 芮达志 陈亚新 李得阳 唐聪智
机构：苏州市泰顺电器有限公司 / 苏州柒整合设计有限公司
国家：中国
组别：产业组

AUTHOR : Zhao Zhifeng, Rui Dazhi, Chen Yaxin, Li Deyang, Tang Congzhi
UNIT : Suzhou Taishun Electric Appliance Co., Ltd. / Suzhou CHY Integration Design Co., Ltd.
COUNTRY : China
GROUP : Product Group

以人为本，注重健康照明已成为业界共识，开始不再只是纯粹地关注光效或使用寿命，而更多地考虑人对光的感觉及光对人的影响，希望能制造出更接近自然光的人造光源，此款台灯光谱非常接近自然光光源。

People-oriented, paying attention to health lighting has become the consensus of the industry, beginning is no longer just pure attention to light effect or service life, and more consider the feeling of light and light to the influence of people, hope to make closer to natural light artificial light source, this desk lamp spectrum is very close to natural light source.

CHECHAL LAMP

竹简铅笔
BAMBOO PENCIL

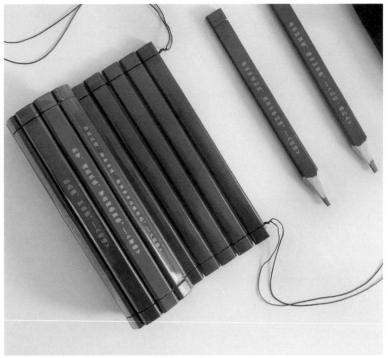

作者：Rodrigo Fernando Diaz Conde
国家：墨西哥
组别：产业组

AUTHOR : Rodrigo Fernando Diaz Conde
COUNTRY : Mexico
GROUP : Product Group

作者：张焱
机构：山东工艺美术学院
国家：中国
组别：产业组

AUTHOR : Zhang Yan
UNIT : Shandong University of Art & Design
COUNTRY : China
GROUP : Product Group

Chechal 灯是一种照明装置，通过仅使用当地材料推进传统作品的形态，并将新产品引入当前市场，从而促进文化创新和手工艺之间的合作。

The chechal lamp is a lighting fixture that promotes cultural innovation and collaboration across artisanal craftmanship, by pushing forward the morphology of their traditional pieces using only endemic materials and introducing new products into a current market.

本产品将优秀文化融入民众生活，以生活化创意实现中华优秀文化的传承与创新。竹简对中国文化的传播起到了至关重要的作用，同时也是孔子时代典型的书写工具。很多儒家经典语句都是从竹书上发掘出来的。以竹简为创意元素，以铅笔为载体，将儒家经典名句印在其上，让人们在使用时能够感受到古老的文字记载方式，也能够感受儒家文化。

This product integrates excellent culture into people's lives, and achieves the inheritance and innovation of Chinese excellent culture through life oriented creativity.Bamboo slips played a crucial role in the dissemination of Chinese culture and were also a typical writing tool during the Confucius era. Many classic Confucian sentences are excavated from bamboo books. Using bamboo slips as creative elements and pencils as carriers. Printing Confucian classics and famous sentences on it allow people to feel the ancient ways of writing and Confucian culture when using it.

5D 抗寒·极地穿越羽绒
5D ANTI-COLD POLAR DOWN JACKET

衡
BALANCE

机构：宁波太平鸟风尚男装有限公司
国家：中国
组别：产业组

UNIT : Ningbo Peacebird Fashion Men's Wear Co., Ltd.
COUNTRY : China
GROUP : Product Group

作者：马珂
机构：徐州白鳍豚文化传播有限公司
国家：中国
组别：产业组

AUTHOR : Ma Ke
UNIT : Xuzhou Baiqitun Culture Communication Co., Ltd.
COUNTRY : China
GROUP : Product Group

这是一款应用于极地环境的抗寒羽绒服，通过创新的布料材质满足极地探险者以及科考人员的需求。

This is a cold-resistance down jacket applied in the polar environment, which aims to satisfy the requirements of polar explorers and scientific researchers via its innovative fabric materials.

"衡"是一款实木与金属结合，利用模块化和连杆的结构方式，解决功能单一、不便携带问题的便携式绘画箱。

"Balance" is a portable paintbox that combines solid wood and metal and uses modular and connecting rod structure to solve the problems of single function and inconvenient carrying.

多功能翻转衣架
MULTI-FUNCTIONAL FLIP HANGER

手电筒
TRILTE

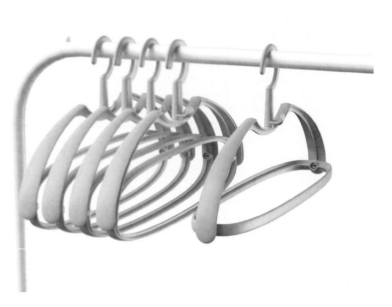

作者：秦夏莹 贾萌 夏祥雷
机构：陕西佳帮手集团控股有限公司 / 西安佳品创意设计有限公司
国家：中国
组别：产业组

AUTHOR : Qin Xiaying, Jia Meng, Xia Xianglei
UNIT : Shaanxi Joybos Group Holdings Co., Ltd. / Xian JIA PIN Creative Design Co., Ltd.
COUNTRY : China
GROUP : Product Group

作者：李卫东 刘玉珍 黄海川 姚晓津 赵乐
机构：深圳市新起点发展有限公司
国家：中国
组别：产业组

AUTHOR : Li Weidong, Liu Yuzhen, Huang Haichuan, Yao Xiaojin, Zhao Le
UNIT : VSTART
COUNTRY : China
GROUP : Product Group

多功能翻转衣架是通过撑开晾晒的方式来解决普通衣架难以将衣物晾干、衣物鼓包问题的产品。

Multi-functional flip hanger is a product that solves the problem of clothings being difficult to dry and bulging by spreading them out for drying.

Trilte 手电筒具有三个可调节角度的光源，聚焦光源的同时增大了光源的照射范围。它的尾盖具有磁吸功能，可以吸附在金属表面进行固定；握把处带有金属夹子，可以借用外物夹住进行固定。产品风格偏向军工风，造型硬朗，可调节角度的光源和丰富的固定方式使得产品更具使用价值，相对于传统的手电筒是一个极大的创新。

Trilte flashlight has three light sources with adjustable angles. Focusing the light source increases the irradiation range of the light source at the same time. Its tail cover has a magnetic suction function and can absorbFixing on the metal surface; There is a metal clip at the grip, which can be clamped by foreign objects for fixation. The product style is biased towards military style, with strong shape and adjustable angle.The high light source and rich fixing methods make the products more valuable, which is a great innovation compared with the traditional flashlight.

气囊式护脊书包
AIR BACKPACK, SPINE PROTECTIVE SCHOOL BAG

作者:冯迪 斯春汝 王伟 郑高伟
机构:宁波天虹文具有限公司 / 浙江工业大学设计与建筑学院
国家:中国
组别:概念组

AUTHOR : Feng Di, Si Chunru, Wang Wei, Zheng Gaowei
UNIT : Ningbo Tianhong Stationery Co., Ltd. / School of Design and Architecture,
Zhejiang University of Technology
COUNTRY : China
GROUP : Concept Group

气囊式护脊书包通过双重气囊背负系统,自适应肩背部身体曲线,并给予运动缓冲,减轻儿童肩背部压力。

Through the Air Cushion System, the Air-type Spine Protecting Schoolbag adopts to the body curve of shoulder and back, and provides buffer to reduce the pressure on children's shoulder and back.

圆规尺
CIRCLE RULER

作者:蔡杨超 李珍妮 胡忠保
机构:杭州布谷工业设计有限公司
国家:中国
组别:产业组

AUTHOR : Cai Yangchao, Li Zhenni, Hu Zhongbao
UNIT : Hangzhou BUGU Industrial Design Co., Ltd.
COUNTRY : China
GROUP : Product Group

圆规尺是一款将画圆和画直线功能结合在一起的产品,解决了学生没有携带圆规或者直尺的问题。

The circle ruler is a way to solve the problem of students not carrying a compass or ruler by combining the functions of drawing a circle and a straight line.

WAKABACHO
WHARF-50-YEAR-OLD REGIONAL BANK
TO THEATRE

九龙腾空
THE NINE DRAGONS ASCENSION

作者 : Masahiro Katsuki, Masashi Ogiwara, Tetsurou Suzuki
机构 : Wakabachokeikaku SIA
国家 : 日本
组别 : 产业组

AUTHOR : Masahiro katsuki, Masashi Ogiwara, Tetsurou Suzuki
UNIT : Wakabachokeikaku SIA
COUNTRY : Japan
GROUP : Product Group

作者 : 罗强明 刘俊聪 金杭杭 代龙 楼晓
机构 : 杭州求索文化科技有限公司
国家 : 中国
组别 : 产业组

AUTHOR : Luo Qiangming, Liu Juncong, Jin Hanghang, Dai Long, Lou Xiao
UNIT : Hangzhou Seeker Culture Technology Co., Ltd.
COUNTRY : China
GROUP : Concept Group

在半个世纪前，这栋建筑是商店协会合作创办的一家银行。在完成这个角色并悄无声息地再现了这个小镇作为各行各业人士聚集的"码头"历史之后，我想要拉开一个新"故事"的帷幕。

This building was a bank that the store association cooperated in half a century ago. After finishing the role and quietly reproducing the history of the town as a "wharf" where cross-border people meet, I would like to open the curtain of a new "story".

作品是以中国古代绘画《九龙图》为灵感，创作的 VR 大空间沉浸式体验项目，用于博物馆等文化场所的巡展。

The work is a VR immersive experience project inspired by the ancient Chinese painting *Jiulongtu*, which is used for exhibition in museums and other cultural sites.

KAMPALA DESIGN WEEK

作者 : Eugene Kavuma, Silas Byakutaaga
机构 : Kampala Design Week
国家 : 乌干达
组别 : 概念组

AUTHOR : Eugene Kavuma, Silas Byakutaaga
UNIT : Kampala Design Week
COUNTRY : Uganda
GROUP : Concept Group

我们的服务利用设计节解决了创意企业家获取知识和市场机会的难题。我们致力于通过技能培训、指导和市场准入机会，提高乌干达创意和设计从业者的生活质量。我们的核心资源是我们与之合作的创意社区，我们也将继续通过我们的知识和技能概念（Kola Design Fellowship 和 Design Garages）来培育这个社区。

Our service solves the challenge of access to knowledge and markets for the creative entrepreneurs using a design festival. We are dedicated to improving the quality of life for creatives and designers in Uganda through skills development, mentorship, and market access opportunities. Our core resource is our community of creatives that we work with. This we will continue nurturing through our knowledge and skills concepts (Kola Design Fellowship and Design Garages)

AI 快闪定制一体机
AI FLASH CUSTOMIZATION MACHINE

作者 : 周磊晶 陈钰蝶 张宇琦 黄银丽 张雨昕
机构 : 杭州设集科技有限公司 / 浙江大学
国家 : 中国
组别 : 产业组

AUTHOR : Zhou Leijing, Chen Yudie, Zhang Yuqi, Huang Yinli, Zhang Yuxin
UNIT : Hangzhou Sheji Technology Co., Ltd. / Zhejiang University
COUNTRY : China
GROUP : Product Group

AI 快闪定制是一款融合人工智能与设计的文创产品一体机，为用户提供个性化、智能化、艺术感的产品定制服务。

AI flash customization is a machine for cultural and creative products that integrates artificial intelligence and design, providing users with personalized, intelligent and artistic product customization services.

青空灯
SKY LIGHT

作者：Byoengchan Oh
机构：Hongik University Graduate School
国家：韩国
组别：概念组

AUTHOR : Byoengchan Oh
UNIT : Hongik University Graduate School
COUNTRY : South Korea
GROUP : Concept Group

青空灯的灵感来自建筑物的天窗，是一种使用透明显示屏的物联网照明装置。

Inspired by skylights in architecture, Sky Light is an IoT lighting that utilizes a transparent display.

连接宝贝凳
CONNECT STOOL

作者：Zhixiang Tao, Xuhui Chen, Mengmeng Wang, Wei Huang, Xuchen Qi
机构：Politecnico di Milano
国家：意大利
组别：概念组

AUTHOR : Zhixiang Tao, Xuhui Chen, Mengmeng Wang, Wei Huang, Xuchen Qi
UNIT : Politecnico di Milano
COUNTRY : Italy
GROUP : Concept Group

幼儿园的孩子们可以从小接触集体生活，是小朋友的快乐天地。根据幼儿园的日常活动设计出连接宝贝凳 CONNECT，CONNECT 具有圆润的造型，特别是赋予了凳子互相连接的功能。孩子们可以随意地把凳子拼接在一起，互相亲近，或者围绕老师一起做游戏，这个过程是孩子们想要的。

The children in kindergarten can touch collective life from childhood, and kindergarten is the happy world of children. CONNECT is designed according to the daily activities of the kindergarten, and CONNECT has a round shape, especially the function to CONNECT the stools with each other. Children can arbitrarily CONNECT the stools together, get close to each other, or play games around teachers, which is a process that children want.

"C-BIRD" 智能儿童鸟类积木玩具
"C-BIRD" INTELLIGENT CHILDREN'S BIRD BUILDING BLOCK TOY

零食收纳盒
SNACK STORAGE BOX

作者：张璐泽 朱家琪 沈亚丽 王可欣 杨婷琳
机构：浙江工业大学
国家：中国
组别：概念组

AUTHOR : Zhang Luze, Zhu Jiaqi, Shen Yali, Wang Kexin, Yang Tinglin
UNIT : Zhejiang University of Technology
COUNTRY : China
GROUP : Concept Group

作者：区锡钊 温建平 赵玉强 江琦
机构：广东意可可科技有限公司
国家：中国
组别：产业组

AUTHOR : Ou Xizhao, Wen Jianping, Zhao Yuqiang, Jiang Qi
UNIT : Guangdong Ecoco Technology Co., Ltd.
COUNTRY : China
GROUP : Product Group

我们的创新在于将鸟类常识与积木有效结合，不同的简化鸟元素带来不同的拼搭方式，灵活的调节设计，可以模拟鸟类不同场景的体态变化，既富有趣味也能激发创造力。每一种拼搭成果借助 RFID 智能识别的方式，与实际的鸟类形成匹配，带来除拼搭玩法之外的惊喜和乐趣，通过在终端上识别显示的鸟类信息，赋予功能单一的积木玩具以鸟类科普教育的意义。

Our innovation is to effectively combine the general knowledge of birds with the blocks, different simplified bird elements bring different ways to put together, flexible adjustment design, can simulate the physical changes of birds in different scenes, both interesting and can stimulate creativity. Each kind of assembling results with the help of RFID intelligent identification, and the actual birds to form a match, bringing surprises and fun in addition to the assembling play, through the identification of the display of bird information on the terminal, giving a single function of the building block toys to bird science popularization significance.

这是一款有手机支架、牙签收纳功能的可叠加式零食收纳盒。
1. 独创手机支架，完美与提手结合，轻轻翻转即可解放双手享受美食；
2. 内设隐藏式牙签收纳盒，吃干果零食时更方便地拿取牙签；
3. 透明可视防尘盖搭配多格分区收纳，合盖时内部情况一目了然，便于拿取；
4. 大容量叠加式收纳，根据需求增加纵向收纳空间，美化桌面，提升空间利用率。

It's a stackable snack storage box with cell phone holder and toothpick storage:
1.Original cell phone holder, perfectly combined with the carrying handle, gently flip to free your hands to enjoy the food. 2.Built-in hidden toothpick storage box, more convenient to pick up toothpicks when enjoying food. 3.Transparent visual dust cover with multi-compartment storage, easy to see the internal situation when the cover is closed. 4. Large-capacity stacking storage, according to demand to increase vertical storage space, beautify the desktop, improve space utilization.

冰箱饮品壶
REFRIGERATOR DRINK CONTAINERS

作者：李科敏 温建平 赵玉强 江琦
机构：广东意可可科技有限公司
国家：中国
组别：产业组

AUTHOR : Li Kemin, Wen Jianping, Zhao Yuqiang, Jiang Qi
UNIT : Guangdong Ecoco Technology Co., Ltd.
COUNTRY : China
GROUP : Product Group

一款用于冰箱的饮品储存水壶。底部斜面设计，不积水，有效避免水资源浪费与健康问题；大口径顶盖配合局部翻盖设计，便于添加饮品；便于安装的拔插式创新型密封水龙头，搭配过滤网，可在自由调节水流大小的同时有效隔绝饮品渣；整体彩色与半透明材料结合，增加其辨识度，便于使用者直观观察壶内饮品。

A beverage storage jug for the refrigerators. Bottom slope design, no water accumulation, effectively avoid water waste and health problems. The large-diameter top cover and the partial flip cove are convenient for adding drinks. The easy-to-install plug-in innovative sealed faucet with a filter, which can effectively isolate drink residues while freely adjusting the water flow. The overall color is combined with translucent materials to increase its recognition and intuitive observation of the drinks in the pot.

悬挂内衣收纳盒
HANGING UNDERWEAR STORAGE BOX

作者：秦夏莹 夏祥雷
机构：陕西佳帮手集团控股有限公司 / 西安佳品创意设计有限公司
国家：中国
组别：产业组

AUTHOR : Qin Xiaying, Xia Xianglei
UNIT : Shaanxi Joybos Group Holdings Co., Ltd. / Xian JIA PIN Creative Design Co., Ltd.
COUNTRY : China
GROUP : Product Group

悬挂内衣收纳盒可粘贴在衣柜内的挂衣区使用，可以节省衣柜内部的空间，产品采用抽拉翻转向下的结构，设计的角度是108°，确保产品拉下来之后衣物不会掉落，方便消费者取内裤、袜子，让寻找内衣的过程简单高效，同时利用了衣柜顶部的闲置空间。外观设计风格简洁干净，注重产品的体验感。申请了1项实用新型专利，5项外观专利。

The hanging underwear storage box is pasted in the clothes hanging area in the wardrobe, which can save the space inside the wardrobe. The product adopts a pulling and turning down mechanism, and the design angle is 108° to ensure that the clothes will not fall after the product is pulled down. It is convenient for consumers to take out underwear and socks, making the process of looking for underwear simple and efficient, and making use of the idle space on the top of the wardrobe. The design style is simple and clean, and the product experience is emphasized. Applied for 1 utility model patent and 5 appearance patents.

皮囊沙发
MASS SOFA

作者：秦帅 党心宇
机构：杭州挑时家居设计有限公司
国家：中国
组别：产业组

AUTHOR : Qin Shuai，Dang Xinyu
UNIT : TELLS STUDIO
COUNTRY : China
GROUP : Product Group

这是我们在2021年的全新艺术项目，关于欲望的思考，与家具的重塑，我们尝试解剖一张沙发，透明的 TPU 和优雅的不锈钢框架，让沙发的内在一览无余，多种多样的填充物，体现了每个使用者的欲望，又从某种意义上重新塑造了沙发本身。造物，造悟，是家具人格化的全新呈现。

This is our brand-new art project in 2021. Thinking about desire, and the reconstruction of furniture, we try to dissect a sofa with transparent TPU & beautiful stainless-steel frame, which let the inside of the sofa be unobstructed, Variety of fillers embodies the desires of every user. In a certain sense, the sofa itself is reconstituted. Create objects, Create perception, a new presentation of furniture personification.

ENPITSU SHARP

作者：Lida Kohei, Fukawa Shohei
机构：KOKUYO Co.,Ltd.
国家：日本
组别：产业组

AUTHOR : Lida Kohei, Fukawa Shohei
UNIT : KOKUYO Co.,Ltd.
COUNTRY : Japan
GROUP : Product Group

Enpitsu Sharp 自动铅笔和铅笔芯采用像铅笔一样的简单设计，并可提供各种铅笔芯尺寸。 我们通过重新审视木制铅笔的优点，重新设计了这款铅笔，旨在打造出与经典木制铅笔一样简单的设计和体验。

Enpitsu Sharp Mechanical Pencils and Leads feature a simple design like a pencil and a wide range of lead sizes. We renewed the design by re-examining the benefits of the wooden pencil, and aimed to create a design and experience as simple as a classic wooden pencil.

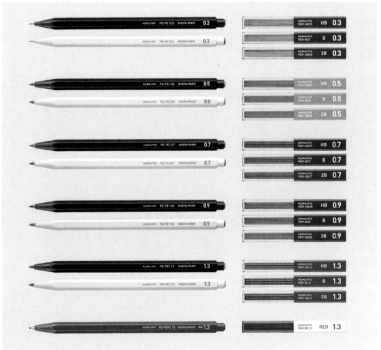

CALZONE

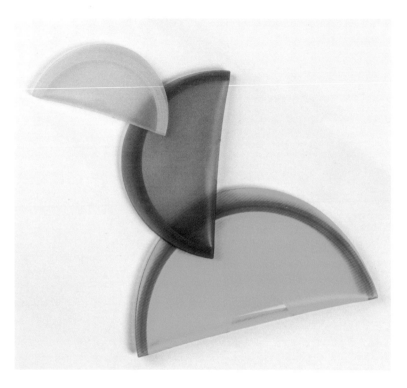

电里拉提提
ELECTRIC NYATITI

作者 : Seomin Lee
机构 : BDCI
国家 : 韩国
组别 : 产业组

AUTHOR : Seomin Lee
UNIT : BDCI
COUNTRY : South Korea
GROUP : Product Group

作者 : Adam Yawe
国家 : 肯尼亚
组别 : 概念组

AUTHOR : Adam Yawe
COUNTRY : Kenya
GROUP : Concept Group

Calzone 是一种可重复使用的折叠盘,设计用于取代一次性产品。对于需要单独盖子或一次性拉链袋的存储容器来说,Calzone 可能是一种新的解决方案。

Calzone is a reusable folding plate designed to replace disposable products. It can be a new solution for storage containers that requires a separate lid or disposable zipper bags.

里拉提提是肯尼亚卢奥部落的一种传统七弦琴。电里拉提提这款产品解决了音乐家希望使用肯尼亚传统乐器里拉提提进行现场表演或录音的问题。

The Nyatiti is a traditional lyre of the Luo tribe. The electric nyatiti is a product that solves the problem of musicians hoping to perform live or record using the traditional Kenyan instrument known as the Nyatiti.

SURPRISE SEED

作者：Leticia Angélica Huelgas Licona
机构：Universidad Autónoma del Estado de México
国家：墨西哥
组别：概念组

AUTHOR : Leticia Angélica Huelgas Licona
UNIT : Universidad Autónoma del Estado de México
COUNTRY : Mexico
GROUP : Concept Group

Surprise Seed 是一款好玩的产品，通过社交设计来解决儿童过度性化的问题，为4至6岁的儿童及其社交直接环境（他们的父母和老师）创造良知。这款产品由一个毛绒玩具和一个电子应用程序组成。

Surprise Seed is a playful product that solves the childhood hypersexualization problem through social design, creating conscience in 4 to 6 year-old children and their social immediate environment: their parents and teachers. It consists of a toy plush and an electronic app.

汉语链 —— 中文学习和汉语文化交流平台
CHINESE CHAIN — A PLATFORM FOR CHINESE LEARNING

作者：张德寅 沈诚仪 成加豪 卢杨 叶琦钧
机构：浙江大学
国家：中国
组别：概念组

AUTHOR : Zhang Deyin, Shen Chengyi, Cheng Jiahao, Lu Yang, Ye Qijun
UNIT : Zhejiang University
COUNTRY : China
GROUP : Concept Group

汉语链是世界上为数不多的针对汉语书写和汉语词卡式学习的语言学习软件。对中文非熟练者极为友好。汉语链立足汉语象形文字的特性，用书写和词卡学习的方式，让学习汉语和使用汉语变得更有趣。这不仅是一个学习软件，也承担着文化交流的任务。汉语链希望汉语学习能够更加有趣，汉语传播能够更加广泛。汉语链能够很好地适配目前市场上所有的智能手机。

Chinese Chain is one of the few language learning software for Chinese writing and Chinese word card learning in the world. Chinese Chain is extremely friendly to those who are not proficient in Chinese. Based on the characteristics of Chinese hieroglyphs, the Chinese Chain uses writing and wordcards to make learning and using Chinese more interesting. This is not only a learning software, but also undertakes the task of cultural exchange. The Chinese Chain hopes that Chinese learning can be more interesting and Chinese communication can be more extensive. Chinese Chain can be well adapted to all smartphones in the current market.

SPECTRA

NYX

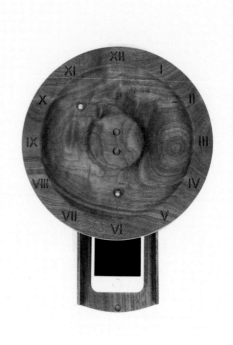

作者 : Matteo Rocchitelli
国家 : 意大利
组别 : 概念组

AUTHOR : Matteo Rocchitelli
COUNTRY : Italy
GROUP : Concept Group

作者 : Dorian Étienne
国家 : 法国
组别 : 概念组

AUTHOR : Dorian Étienne
COUNTRY : France
GROUP : Concept Group

该项目旨在为人与植物之间的关系提供一个新的视角，摒弃以人为中心的概念，转而采用后人类视角。为了实现这一目标，开发了"Spectra"，这是一种照明系统，可以共存于单一产品中，满足新工人和工厂的照明和福祉需求。通过读取电位差，可以开发出一种生物动力照明系统，该系统能够在单个产品中在人和植物之间积极互动，能够满足两个参与者的光需求。

The project aims to offer a new vision of the relationship between man and plant, abandoning the concept of human-centered in favor of a post-human vision. In response to this goal, "Spectra" has been developed, a lighting system to coexist in a single product, which can meet the needs of lighting and well-being of new workers and plants. Through the reading of the difference in electrical potential, it was therefore possible to develop a biodynamic lighting system capable of actively interacting between man and plant in a single product, capable of satisfying the light needs of both actors.

这种交互式睡眠对象有助于发现并遵循我们的生物睡眠周期，从而实现最佳的长期睡眠，以获得更好的生活。

This interactive sleep object helps to find and follow our biological sleep cycles, thereby allowing optimal long-term sleep for a better life.

DOT

MEMORABLE

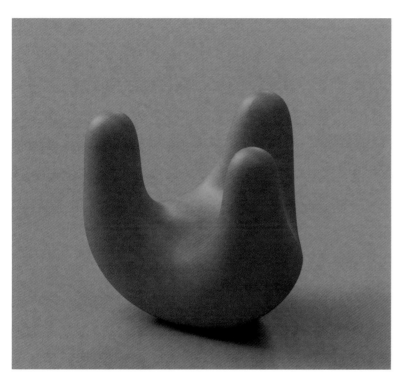

作者 : Daniela Grisel Beizaga
机构 : IED Madrid
国家 : 西班牙
组别 : 概念组

AUTHOR : Daniela Grisel Beizaga
UNIT : IED Madrid
COUNTRY : Spain
GROUP : Concept Group

作者 : Tuncay ince
国家 : 土耳其
组别 : 概念组

AUTHOR : Tuncay ince
COUNTRY : Turkey
GROUP : Concept Group

这项服务利用设计节为创意企业家解决获取知识和进入市场的难题。设计师的核心资源是与我们合作的创意社区。他们将通过我们的知识和技能理念（科拉设计奖学金和设计车库）继续得到培养。设计师对文化产品的消费进行标准化和货币化，为创造这些产品的创意和文化生产者带来价值和增长。

The service solves the challenge of access to knowledge and markets for the creative entrepreneur using a design festival.The designer's core resource is our community of creatives that we work with. They will continue nurturing through our knowledge and skills concepts (Kola Design Fellowship and Design Garages). The designers standardize and monetize the consumption of cultural outputs, adding value and growth to the creative and cultural producers who generate these outputs.

MEMORABLE 是一种卷笔刀设计，通过将削下的铅笔屑困在里面，让透明的 TPU 面料制成的字母变成一个配件（物件）。

MEMORABLE is a sharpener design that allows letters made of transparent TPU fabric to turn into an accessory (object) by trapping pen waste inside.

NIVOO | ADULT VERSION

长颈鹿之光
TWIGA MWANGAZA

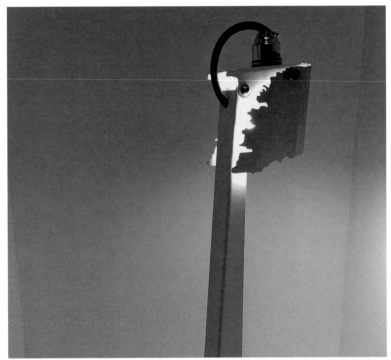

作者 : Christian Kroepfl, Julian Wudy, Alois Füchsl, Colin Vickers
机构 : Guut GmbH / Christian Kroepfl Architecture & Design
国家 : 奥地利
组别 : 产业组

AUTHOR : Christian Kroepfl, Julian Wudy, Alois Füchsl, Colin Vickers
UNIT : Guut GmbH / Christian Kroepfl Architecture & Design
COUNTRY : Austria
GROUP : Product Group

作者 : Kashyap Pravin Gohel
机构 : Kenya Lighting Industries
国家 : 肯尼亚
组别 : 产业组

AUTHOR : Kashyap Pravin Gohel
UNIT : Kenya Lighting Industries
COUNTRY : Kenya
GROUP : Product Group

我们的基本任务是开发一款可持续、无金属、可大规模生产和可模块化扩展的办公桌，以补充现有的家具系列。由于需求量大，目前正在开发成人款。

The basic task was to develop a sustainable, metal-free, mass-produced and modularly expandable desk that could complement the existing furniture line. Due to high demand, an adult version is currently developed.

Twiga Mwangaza，从斯瓦希里语翻译过来就是"长颈鹿之光"的意思，是对一只非洲长颈鹿咀嚼金合欢树叶的优雅抽象，变成用于工作照明的可调节台灯或用于客厅和卧室的环境照明灯和阅读灯。

Twiga Mwangaza, translated from Kiswahili as'Giraffe Light'is an elegant abstraction of an African Giraffe munching on Acacia tree leaves, into an adjustable desk lamp for task lighting or ambient lighting and reading lamps for living rooms & bedrooms.

BAMBOO PRODUCTS

作者：Charles Njoroge, Sharon Wanjiku, Mutura Kuria, Rose Kamau, Michael Mbugua
机构：Opulent Kenya
国家：肯尼亚
组别：产业组

AUTHOR : Charles Njoroge, Sharon Wanjiku, Mutura Kuria, Rose Kamau, Michael Mbugua
UNIT : Opulent Kenya
COUNTRY : Kenya
GROUP : Product Group

我们为客户提供多样化的高品质竹制品。我们与该地区的多家公司合作，为他们提供竹制品，并通过零售合作伙伴推广我们的品牌。我们与建筑师、设计师和创意人员合作，用竹子来满足小众需求。

We provide our customers with a diverse range of high quality bamboo products. We partner with a diverse range of companies in the region, supplying them with bamboo products and promoting our brand through retail partners. We work with Architects, designers and creatives to fulfill niche needs with bamboo.

KENYA BAMBOO FESTIVAL

作者：Charles Njoroge, Sharon Muigai, Mutura Kuria, Rose Kamau, Michael Mbugua
机构：Opulent Kenya
国家：肯尼亚
组别：概念组

AUTHOR : Charles Njoroge, Sharon Muigai, Mutura Kuria, Rose Kamau, Michael Mbugua
UNIT : Opulent Kenya
COUNTRY : Kenya
GROUP : Concept Group

我们的目标是举办一个基于竹子功效的节日，以及竹子如何有助于理解各种产品生产中的创新和创造力过程：手工艺品、珠宝、家具和建筑材料。这将能够将艺术家、建筑师、设计师和企业家与公众联系起来。

We aim to host a festival based on the efficacy of bamboo and how it contributes to understanding processes of innovation and creativity in the production of various products: handicrafts, jewelry, furniture and construction material. This will be able to connect artists, architects, designers and entrepreneurs to the public.

FLEXION

作者： Bulent Unal, Elif Gunes
机构： Atilim University
国家： 土耳其
组别： 概念组

AUTHOR : Bulent Unal, Elif Gunes
UNIT : Atilim University
COUNTRY : Turkey
GROUP : Concept Group

我们的场景经常是千篇一律的、熟悉的、传统的，没有什么创新之处。这个设计传达了主角和他们对生活的思考，就像来自未来的兴奋感，与传统生活场景以及我们的生活场景都截然不同。

Our scenes are always the same and familiar, traditional and places with little innovation. This design conveys the leading roles and the reflections of their lives, just like an excitement from the future, with a touch apart from the traditional to our life scene.

桌上森林系列 —— 立体纸艺术植物贴纸
DESKTOP FOREST — 3D PAPER ART PLANT STICKER

作者： 黄文翰
机构： JOE WONG DESIGN CO.
国家： 中国（香港）
组别： 产业组

AUTHOR : Huang Wenhan
UNIT : JOE WONG DESIGN CO.
COUNTRY : Hong Kong SAR, China
GROUP : Product Group

本产品不需要水、阳光、 肥料与土壤，但可以保持翠绿；可以在任何地方随处"种植"，将自然带入工作和生活。

Water, sunlight, fertilizer and soil are not required, but they remain lush and green all the time. Anyone can easily bring nature to anyplace in work and life.

FREEDOM — FLEXIBLE ARCHITECTURAL LIGHT

多合一支撑架
ALL IN ONE CRADLE

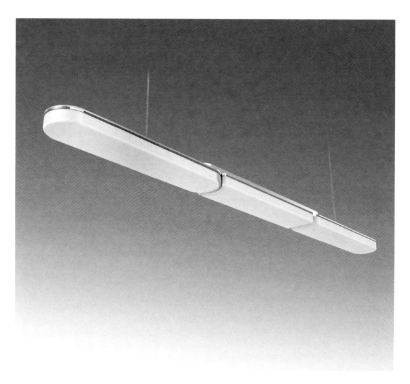

作者：Sumit Singh, Bibhas Ranjan Sethi, Deep Singh, Ashish Chaturvedi
机构：Havells India Ltd.
国家：印度
组别：产业组

AUTHOR : Sumit Singh, Bibhas Ranjan Sethi, Deep Singh, Ashish Chaturvedi
UNIT : Havells India Ltd.
COUNTRY : India
GROUP : Product Group

作者：金完燮
机构：佛山市完完工业设计有限公司
国家：中国
组别：产业组

AUTHOR : Jin Wanxie
UNIT : Foshan Wanwan Industrial Design Co., Ltd.
COUNTRY : China
GROUP : Product Group

FREEDOM 是一款智能、灵活的建筑灯具，旨在适应各种生态系统，并与以人为中心的照明相结合。其设计目的是实现灵活性、模块化和连续性，使建筑师能够自由地设计出无限长度的各种可能形状。

FREEDOM is an intelligent, flexible architectural luminaire designed to adapt varied ecosystem and is integrated with Human Centric Lighting. The design intent is to achieve flexibility, modularity and continuity, giving freedom to the architect to design various possible shapes for infinite length.

本产品为应用普通键盘环境中三种角度，即6°、8°、10°，能够感受低角度舒适感的笔记本电脑、平板电脑多合一支架。

Using the three angles familiar to the use environment of ordinary keyboard, i.e., 6 degrees, 8 degrees and 10 degrees, the laptop and tablet PC which can feel the comfort of low angle are all in one rack.

机器岛儿童学习桌
JIQIDAO CHILDREN'S LEARNING TABLE

作者：吕继伦 翟见文
机构：南京机器岛智能科技有限公司
国家：中国
组别：产业组

AUTHOR : Lv Jilun, Zhai Jianwen
UNIT : Nanjing Jiqidao Intelligent Technology Co., Ltd.
COUNTRY : China
GROUP : Product Group

通过高度可调节的桌面，机器岛学习桌可满足3~18岁不同年龄段孩子的日常使用需求。

Through the height adjustable table, JIQIDAO learning table can meet the daily use needs of children aged 3-18 years old.

OFFICE STRETCHER

作者：Xinyue Guo
国家：美国
组别：产业组

AUTHOR :Xinyue Guo
COUNTRY : United States
GROUP : Product Group

"Office Stretcher"是一个模块化的交互式装置，旨在创建灵活、鼓舞人心、协作和健康的工作空间。

"Office Stretcher" is a modular and interactive installation that is designed to create flexible, inspiring, collaborative, and healthy workspaces.

DIDOO 原木玩具汽车
DIDOO WOOD TOY CAR

作者：张飞 周倩 崔杨彬 汤佳 毕顺
机构：浙江本来家居科技有限公司 / 杭州本来工业设计有限公司
国家：中国
组别：产业组

AUTHOR : Zhang Fei, Zhou Qian, Cui Yangbin, Tang Jia, Bi Shun
UNIT : belaDESIGN wood / belaDESIGN
COUNTRY : China
GROUP : Product Group

嘀嘟汽车是通过基础形态研究对原木加工而设计制作的一系列原木玩具汽车。

Didoo car is a series of wood toy cars designed and manufactured by processing wood through basic morphological research.

趣味八段锦叠叠乐
BUILDING BLOCK TOY OF THE EIGHT BROCADES

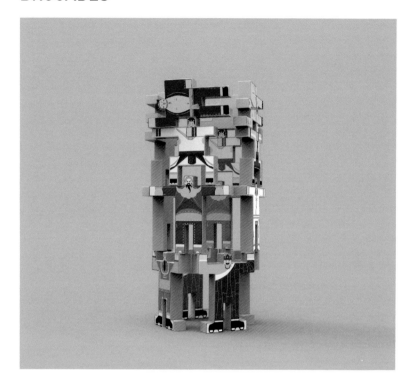

作者：陆明朗 魏婷
机构：宁波美乐雅荧光科技股份有限公司 / 厦门市拙雅科技有限公司
国家：中国
组别：产业组

AUTHOR : Lu Minglang , Wei Ting
UNIT : Ningbo Merryart Glow-tech Co.,Ltd. / Xia Men Joyatech Co., Ltd.
COUNTRY : China
GROUP : Product Group

本产品是以中国传统健身动作八段锦为创意的文化益智积木玩具，让孩子在多样玩法中体验传统文化的趣味性。

This product is a cultural and educational building block toy designed for children over 3 years old. It takes Baduanjin, a traditional Chinese fitness movement, as the creative idea, so that children can experience the fun of traditional culture in a variety of ways.

SECOND SKINS

作者：Malou Beemer, Christian Dils
机构：Malou Beemer, Atelier Mlou
国家：荷兰
组别：概念组

AUTHOR : Malou Beemer, Christian Dils
UNIT : Malou Beemer, Atelier Mlou
COUNTRY : Dutch
GROUP : Concept Group

适合未来健康、可持续衣橱的护理服装。在一个日益智能化的世界里，是时候让我们的衣橱开始适应我们的需求了。在这个跨学科的研究项目中，我们在探索适应我们实际和社会需求的服装。

The adaptive caring garments of our future healthy and sustainable wardrobe .In a world that is becoming smarter every day, it is time that our wardrobe starts to adapt to our needs. Within this interdisciplinary research project, we explore garments that adapt to our practical and social needs.

IICONIC 鞋
IICONIC FOOTWEAR

作者：Iris Camps
机构：Eindhoven University of Technology
国家：荷兰
组别：概念组

AUTHOR : Iris Camps
UNIT : Eindhoven University of Technology
COUNTRY : Dutch
GROUP : Concept Group

IICONIC 鞋是通过3D 打印和环形编织的循环生产流程制成的。它着眼于解决鞋类生产中对环境最有害的阶段，并消除了对原材料和有毒胶水的需求。最终创造出可以在新鞋中复制的鞋子。

IICONIC Footwear is created through a circular production process that includes 3D printing and circular knitting. It looks at the most harmful stages of footwear production for the environment and removes the need for raw materials and toxic glues. Ultimately, creating shoes that can be reprinted in new shoes.

JUCK

作者：Yongil Lee, Jaemyung Seo, Bonghee Kim
机构：Artworks Group
国家：韩国
组别：产业组

AUTHOR : Yongil Lee, Jaemyung Seo, Bonghee Kim
UNIT : Artworks Group
COUNTRY : South Korea
GROUP : Product Group

"Juck"是为了缓解护身符的负面形象，借用护身符作为其设计理念，并从文化角度进行处理。重点是"Juck"的意义传递体系和神力表达。首先，艺术家通过考察迄今为止收集的各种护身符的文字和符号，分析了护身符的意义传递系统。然后，视觉元素通过印刷术表达，扩展到认知领域。

The submitted work, "Juck", was conceived to alleviate the negative image of the amulet by borrowing the amulet as its design concept and approaching it from a cultural perspective. The focus was on the delivery system of meaning and the expression of supernatural powers of "Juck" First, the artist analyzed the delivery system of meaning of amulets by examining the characters and symbols of various amulets collected so far. Then, the visual elements were expanded into the sphere of cognition by expressing them via typography.

FLIPPACK 防水防盗折叠背包
FLIPPACK

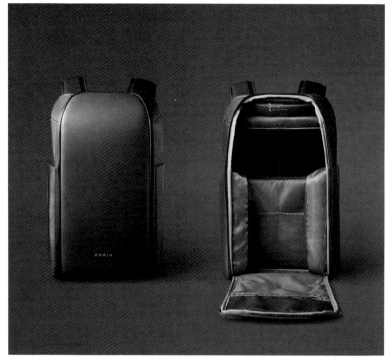

作者：黄辉古 赵璧 曾繁渠
机构：广州市阔云科技有限公司
国家：中国
组别：产业组

AUTHOR : Huang Huigu, Zhao Bi, Zeng Fanqu
UNIT : Guangzhou Korin Technology Co., Ltd.
COUNTRY : China
GROUP : Product Group

FlipPack 是一款针对城市出行的轻科技功能背包。随着人们短途出行的频次不断增加，市场需求的迅速增长和消费者需求趋向个性化，针对出行设计一款立体防盗背包是市场所需。FlipPack 使用 TSA 海关锁、防盗绳、磁吸独立仓等，彻底解决用户城市出行的各种有关防盗、寄放、减负、收纳的行为与心理痛点。FlipPack 以设计思维将新技术植入箱包中，重新审视人们的生活方式、使用需求，并进行功能性创新，为轻科技功能箱包带来发展机会。

FlipPack is a light technology functional backpack for urban travel. With the increasing frequency of short-distance travel, the rapid growth of market demand and the tendency of consumer demand to personalization, designing a stiff anti-theft backpack for travel is what the market needs.FilpPack uses TSA customs locks, anti-theft cords, magnetic independent compartments, etc. to thoroughly solve various behavioral and psychological pain points of users related to anti-theft, consignment, load reduction and storage for urban travel.FlipPack takes design thinking to implant new technologies into luggage, re-examine people's lifestyles, use needs, and make functional innovations, bringing development opportunities for light technology functional luggage.

净酌
JO-CHU

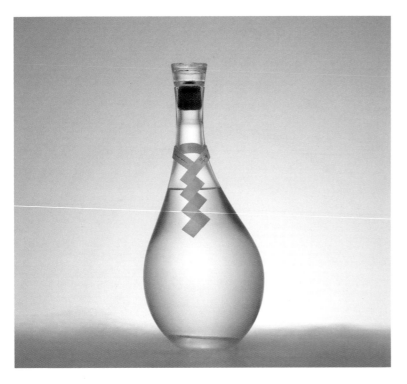

作者： Eisuke Tachikawa, Ryota Mizusako, Jin Nagao, Daichi Komatsu, Naoki Hijikata
机构： NAORAI Co., Ltd. NOSIGNER
国家： 日本
组别： 产业组

AUTHOR : Eisuke Tachikawa, Ryota Mizusako, Jin Nagao, Daichi Komatsu, Naoki Hijikata
UNIT : NAORAI Co., Ltd. NOSIGNER
COUNTRY : Japan
GROUP : Product Group

JO-CHU（净酌）是一种新的日本酒，它通过前所未有的低温蒸馏工艺，继承了清酒的风味，或者可以说是灵魂。这是一个将清酒文化融入未来的品牌，在其设计中体现出其精神性。

JO-CHU is a new type of Japanese alcohol that inherits the flavor, or soul, so to speak, of sake through an unprecedented process of low-temperature distillation.It is a brand that weaves the sake culture into the future, expressing its spirituality in its design.

相见欢温酒器
XIANG JIAN HUAN WINE WARMER

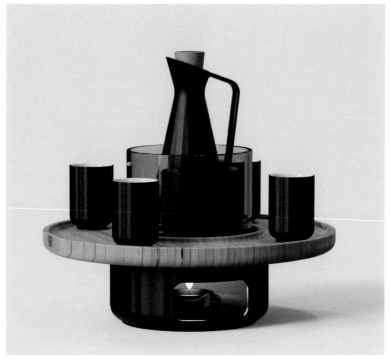

作者： 郑润桦 黄俊杰 兰鼎平
机构： 南山先生（厦门）文化创意有限公司
国家： 中国
组别： 产业组

AUTHOR : Zheng Runhua, Huang Junjie, Lan Dingping
UNIT : Mr. Nan Shan (Xiamen) Cultural Creativity Co., Ltd.
COUNTRY : China
GROUP : Product Group

相见欢温酒器造型风格上采用了现代简约的条纹元素，把手的设计让产品更好抓持，长直壶嘴的设计让断水更加利落顺畅，竹制圆盘既可以拿下来当托盘，也可以架在底部暖炉上使用；材质上进行了跨材质的组合搭配，由瓷胎陶釉、有色高硼硅玻璃、竹质三种材质组合而成，陶瓷的厚重质感，玻璃的通透、轻盈感，竹子的亲和力和温和感、被巧妙地结合在一起。

With the stripe elements, the modeling style of Xiangjianhuan wine warmer adopts modern and reductive design. The design of the handle makes it easy to grasp while it is very convenient to pour the wine with a long straight spout. The bamboo circular tray can be used as a tray or put above the bottom stove. As for the materials, it is made from porcelain and ceramic glaze, colored high-borosilicate glass as well as bamboo, which is an ingenious combination for the dignified texture of ceramics, the lightness of glass, as well as the gentleness of bamboo.

传家壶 —— 迷你
HEIRLOOM THERMOS — MINI

作者：季坤荣 夏飞剑 孔洪强 刘安民 吴冬
机构：深圳市一原科技有限公司
国家：中国
组别：产业组

AUTHOR : Ji Kunrong, Xia Feijian, Kong Hongqiang, Liu Anmin, Wu Dong
UNIT : ShenZhen Yiyuan Technology Co., Ltd.
COUNTRY : China
GROUP : Product Group

传家壶 —— 迷你是一款将现代科技与传统工艺结合，具有东方美学气质，适合当代生活场景的保温壶。

Heirloom thermos-Mini is a modern technology and traditional technology combined with Oriental aesthetic temperament, suitable for the contemporary life scene of the thermal insulation pot.

雅琮提梁壶
YA CONG HANDLE POT

作者：徐乐 包力源 辜弯婉 陈欣 沈堃昊
机构：杭州大巧创意设计有限公司 / 浙江工业大学之江学院
国家：中国
组别：产业组

AUTHOR : Xu Le, Bao Liyuan, Gu Wanwan, Chen Xin, Shen Kunhao
UNIT : Hangzhou Great Wisdom Creative Design Co., Ltd. / Zhijiang College of Zhejiang University of Technology
COUNTRY : China
GROUP : Product Group

雅琮提梁壶是一款良渚玉琮文化与紫砂艺术相结合的茶器。运用设计的创新方法将玉琮造型运用在紫砂壶作品的创作中，整体造型以"外方内圆"的特征来进行系统化设计，强调刚柔并济；提梁与壶身采用暗接的手法进行衔接，表达出壶的整体感和气韵相通。产品造型古朴简约大气，具有较高的艺术价值，又具有实用功能。

YA CONG Handle Pot is a tea vessel combining Liangzhu jade cong culture and Zisha art.The overall shape is systematically designed with the characteristic of "square outside and round inside", emphasizing both rigidity and softness; the lifting beam and the body of the pot are articulated with the technique of concealed connection, expressing the overall sense of the pot and the harmony of qi. The shape is simple, simple and atmospheric, with high artistic value and practical function.

SOFTLINE

消波块 模块家具
MAP STOOL

作者：Ahmet Osman PEKER
机构：Kar porselen
国家：土耳其
组别：产业组

AUTHOR : Ahmet Osman PEKER
UNIT : Kar porselen
COUNTRY : Turkey
GROUP : Product Group

作者：PD A, PD B, PD C
机构：GYRO, PARTISAN 404
国家：中国
组别：产业组

AUTHOR : PD A, PD B, PD C
UNIT : GYRO, PARTISAN 404
COUNTRY : China
GROUP : Product Group

在享受浓郁的咖啡风味和纯粹的瓷器触感的同时，您还可以用应有的仁慈和关怀来对待大自然 ……

While enjoying the rich flavors of coffee enhanced with a pure touch of porcelain, you can also treat nature with the kindness and care it deserves…

最初是为浦东美术馆儿童教育区设计，以单体满足不同年龄观众使用的简易公共家具。概念来自黄浦江堤岸的消波块，采用与江河漂浮垃圾同材质的回收 PE，成品轻巧耐用，安全且易于维护。居家使用时，也解决了儿童家具通常跟不上孩子身体和审美快速成长而带来的弃置浪费，在环保材质外，以功能和设计带来真正的长效性。

Originally designed for the children's educational area at the Museum of Art Pudong, MAP stool is a simple public furniture for audiences of varying ages. The concept is driven by breakwater units on the Huangpu River embankment. Made of recycled PE, which comes from the floating plastic garbage in the river, the product is light and flexible, durable in use, yet safe and easy to maintain. When used at home, it also solves the problem of disposal and waste caused by children's furniture, usually failing to keep up with the rapid growth of children's bodies and aesthetics. In addition to environmentally friendly materials, it brings true sustainability through long lifespan with function and design.

珐琅铁铸饭煲
ENAMELLED IRON CAST RICE COOKER

一席
PRIVATE TEA SPACE

作者：张必锋 杨扬 杨能鹏 卢传德 罗玮瑜
机构：广东顺德米壳工业设计有限公司
国家：中国
组别：概念组

AUTHOR : Zhang Bifeng, Yang Yang, Yang Nengpeng, Lu Chuande, Luo Weiyu
UNIT : MIKO Industrial Design Co.,Ltd.
COUNTRY : China
GROUP : Concept Group

作者：磨炼 王兆巧 陈雯
机构：深圳市聿上家具设计有限公司 / 广州美术学院
国家：中国
组别：概念组

AUTHOR : Mo Lian, Wang Zhaoqiao, Chen Wen
UNIT : YOSON Design Co., Ltd. / Guangzhou Academy of Fine Arts (GAFA)
COUNTRY : China
GROUP : Concept Group

这是一款专为追求高品质精致生活而打造的珐琅铁铸饭煲。产品应用远红外线技术，还原传统柴火饭技法，米饭受热均匀通透，引出米饭原本的香味，符合健康的饮食理念；内部设有独立探温系统，实时监控锅内温度，避免温度过高；创新分段式的设计，顶部为锅盖，中部为铁铸饭锅主体，下部为热传导锅底。

This is an enamel iron cast rice pot specially made for the pursuit of high quality and exquisite life.The product uses far infrared technology to restore the traditional firewood rice technique. The rice is heated evenly and fully, leading to the original aroma of rice, which is in line with the healthy diet concept. An independent temperature detection system is set inside to monitor the temperature of the pot in real time to avoid high temperature; Innovative segmented design, the top is the pot cover, the middle is the iron cast rice pot body, the bottom of the heat conduction pot.

一个可以在居住空间中任意移动的茶盒，满足人们根据自己的心情和喜好切换喝茶环境的需求。

A tea box can be moved freely in the living space to meet people's needs to change the drinking environment according to their mood and preferences.

PLATO COLLECTION

KONI

作者：Kerem Aris, Merve Parnas, Defne Koz
机构：Uniqka
国家：土耳其
组别：产业组

AUTHOR : Kerem Aris, Merve Parnas, Defne Koz
UNIT : Uniqka
COUNTRY : Turkey
GROUP : Product Group

作者：Kerem Aris, Merve Parnas, Romy Kühne
机构：Uniqka
国家：土耳其
组别：产业组

AUTHOR : Kerem Aris, Merve Parnas, Romy Kühne
UNIT : Uniqka
COUNTRY : Turkey
GROUP : Product Group

Plato 托盘专为平静、精致和幸福地整理日常物品而设计。Plato 定义了一个没有限制的微妙界限，对日常生活的无限选择进行组织整理。Plato 在与周围环境融为一体并变得几乎看不见的同时，通过减少杂乱来发表和平声明。

Plato trays are designed for organizing everyday objects with serenity, sophistication and bliss. Plato defines a subtle limit with no limits, organizing the infinite options of daily life. While blending in with its surroundings and becoming almost invisible, Plato is there to make a peaceful statement by reducing clutter.

Koni 是一个完美的例子，展示了数学、几何和物理如何与工艺一起用于产品设计。设计出的几何图案由一块皮革制成灯罩。最终产品的计算和试验耗时两年多，直到 Uniqka 和 Romy Kühne 获得最佳结果。

Koni is a perfect example of how mathematics, geometry and physics are used in the product design together with craftsmanship. The devised geometric pattern creates a lampshade from one single piece of leather. The calculations and trials to reach the final product took over two years, until Uniqka and Romy Kühne had the best result.

戏出东方
SHADOW PLAY IN CHINA

作者：张书雁 王猛涛
机构：浙江自然造物文化创意有限公司
国家：中国
组别：产业组

AUTHOR : Zhang Shuyan, Wang Mengtao
UNIT : Made in nature
COUNTRY : China
GROUP : Product Group

戏出东方是一款礼盒，灵感提取于古老技艺"皮影戏"，初衷是为了重现皮影及让大家去感受和接触皮影文化。

Shadow play in China is a gift box, inspired by the ancient technique of "shadow puppetry". The original intention is to reproduce shadow puppetry and let everyone feel and contact the shadow puppet culture.

搭扣
BUCKLE

作者：姜军 金炜楠
机构：上海瑞鹊投资有限公司 / 杭州乐造工业设计有限公司
国家：中国
组别：产业组

AUTHOR : Jiang Jun, Jin Weinan
UNIT : Shanghai RUIQUE Investment Co., Ltd. / Hangzhou LeDesign Co., Ltd.
COUNTRY : China
GROUP : Product Group

"Buckle"是一款防烫便当盒，聚焦国人饮食场景，通过智巧的设计解决中国人饮食过程中的饭菜加热防烫等问题。

"Buckle" is an anti-hot bento box that focuses on the Chinese food scene and solves problems like heating and anti-hot food in the Chinese eating process through smart design.

玲珑架
LINGLONG SHELF

M+

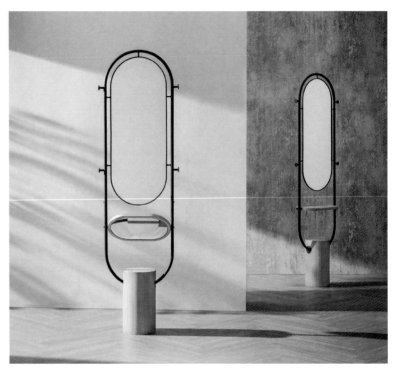

作者：周安彬
机构：深圳得闲设计有限公司
国家：中国
组别：产业组

AUTHOR : Zhou Anbin
UNIT : Shenzhen DeXian Design Co., Ltd.
COUNTRY : China
GROUP : Product Group

作者：薛平安 李超阳
机构：致欧家居科技股份有限公司
国家：中国
组别：产业组

AUTHOR : Xue Pingan, Li Chaoyang
UNIT : Ziel Home Furnishing Technology Co., Ltd.
COUNTRY : China
GROUP : Product Group

玲珑架在这个看似简单的外表下隐藏精巧的设计方式以及对工艺的极致要求，全拆装的结构却没有一颗木榫与钉子，完全依赖于层板开孔的斜度以及立柱斜度的完美配合。三个不同方向的立柱则起到减少左右摆动的关键作用，从上至下逐渐加厚每层的层板，也起到了稳定的作用，并与塔柱的视觉效果相呼应。

LingLong shelf hides the exquisite design method and the extreme request to the craft in this seemingly simple appearance. However, there is no wooden tenon or nail in the whole disassembly structure, which completely depends on the perfect matching with the slope of the opening of the laminate and the inclination of the column. When the laminate is inserted into the column from top to bottom, it is fixed and clamped when the laminate reaches the fully fit position.The three columns with different orientations play a key role in reducing the left and right swing stress, and the gradually thickened laminates in each layer also play a stable role and balance with the tower column visual effect.

M+ 整体造型简约，符合现代审美，基础几何造型的挂钩和上半部的镜子集合了衣帽架和穿衣镜的功能，中间的托盘，上面置物，底部嵌入镜子，可以放置钥匙、手表、钱包、眼镜等小物品，当人们出门时旋转托盘（镜子）与上面的镜子平齐，形成大的全身镜。以用户的生活习惯和简约原则为出发点，作为一站式解决方案为生活提供一些便利。

Featuring a clean, simple look, M+ Mirror oozes a modern touch. The hooks in geometric shape and the upper mirror works as a coat rack with a mirror. The storage tray in the middle offers a perfect perch for keys, watches, and glasses. With an adjustable design and a mirror on the bottom side, the tray doubles as a complementary mirror. Adjust the mirror side flush with the upper mirror for a large, full-length mirror.Based on people's living habits and the principle of "less is more", M+ Mirror provides a one-stop solution for people to get ready for the day, with ease.

PER 系列
PER SERIES

作者： Miyu Ikeda , Takuto Kurashima
机构： Hirata Chair Manufacture Co., Ltd.
国家： 日本
组别： 产业组

AUTHOR : Miyu Ikeda , Takuto Kurashima
UNIT : Hirata Chair Manufacture Co., Ltd.
COUNTRY : Japan
GROUP : Product Group

"Per 系列"是专为居住在小房子里的年轻人设计的一系列紧凑型家具产品。该系列包含8件产品：凳子、椅子、2种桌子和2种长凳。

The "Per Series" is a series of compact furniture products designed for young people living in small houses. The collection consists of 8 items: stool, chair, 2 types of tables, and 2 types of benches.

MANA

作者： Ozan Tığlıoğlu , Wakako Esra Aras , Furkan Öz
机构： GOA Design Factory
国家： 土耳其
组别： 产业组

AUTHOR : Ozan Tığlıoğlu , Wakako Esra Aras , Furkan Öz
UNIT : GOA Design Factory
COUNTRY : Turkey
GROUP : Product Group

在公共空间创造良好、可持续的声音环境的解决方案变得越来越重要。与生活、自然环境的魔力联系可以被视为对减少噪声和恢复平静的日益增长的需求的回应，例如在办公室、餐厅和其他会议场所。在为人们定义一个特殊区域的同时，我们设计MANA产品的方式可以捕捉人的本质，并用自然的线条与空间连接，而不是用苛刻的设计语言。

Solutions for creating good, sustainable sound environments in public spaces are becoming more and more important. The MANA connection to living, natural environment can be seen as a response to the growing need to reduce noise and restore calmness, for example in offices, restaurants and other meeting places. While defining a special area for people, we have designed the MANA product in a way that can capture the essence of people and connect with space with natural lines, not with a harsh design language.

熊抱
BEAR HUG

INFINITE GROWTH

作者：缪景怡 林丹慧 尹雪
国家：中国
组别：产业组

AUTHOR : Miao Jingyi, Lin Danhui, Yin Xue
COUNTRY : China
GROUP : Product Group

作者：Jun Wang, Xia Hua, Lingxiao Zeng, Yining Xu, Mingdong Mao
机构：Junwang Studio
国家：意大利
组别：概念组

AUTHOR : Jun Wang, Xia Hua, Lingxiao Zeng, Yining Xu, Mingdong Mao
UNIT : Junwang Studio
COUNTRY : Italy
GROUP : Concept Group

"生活派"在居家办公时追求舒适、趣味的办公方式。熊抱是一款可以变换不同姿势来缓解办公压力的家具，且当用户需要进行床上办公时，可以拆分坐具的互补形体，将上半部分羊毛坐具放在床上进行办公。不同用户可以调整摇椅底部的小球，使摇椅重心符合自己的晃动幅度。采用平板化包装设计，可节省70%的运输空间。

People are pursuing a comfortable and interesting way of working. Bear Hug is a furniture that can change different positions to relieve office pressure. When users need to work in bed, they can split the complementary shape of the seat and put the upper part of the wool seat on the bed for the office. Different users can adjust the ball at the bottom of the rocking chair to make the center of gravity of the rocking chair conform to their shaking range. The flat packaging design can save 70% of the transportation space.

这是一件鼓励孩子和父母一起组装的家具，增加了家庭中的互动。同时，这件家具本身具有多种属性，可满足儿童成长不同阶段对家具的需求。

It is a piece of furniture that encourages children and their parents to assemble together, which increases the interaction in the family. At the same time, the furniture itself has a variety of attributes to meet the needs of furniture at different stages of children's growth.

这里有猫腻
THERE IS SOMETHING GOING ON

作者：吴曲
国家：中国
组别：产业组

AUTHOR : Wu Qu
COUNTRY : China
GROUP : Product Group

这是专为爱猫人士设计的一款多用途家具，人宠共用是其独特的产品特点。

It is specially designed for cat lovers, a multi-purpose furniture, pet sharing as the product's unique product features.

铣型椅
CNC

作者：Yrjo Kukkapuro
机构：上海阿旺特家具有限公司
国家：中国
组别：产业组

AUTHOR : Yrjo Kukkapuro
UNIT : Shanghai Avarte Furniture Co.,Ltd.
COUNTRY : China
GROUP : Product Group

CNC椅运用CNC设备通过铣型方式解决了座椅对于复杂结构和复杂座椅机构的需求，从而达到设计与艺术的要求。

CNC chairs are produced through CNC equipment through milling methods to solve the chair's needs for complex structure and complex seat mechanism, to achieve the requirements of design and art.

BAN

RELICS

作者：Yong-il Lee, Jaemyeong Seo, Bonghee Kim
机构：Minjakso
国家：韩国
组别：产业组

AUTHOR : Yong-il Lee, Jaemyeong Seo, Bonghee Kim
UNIT : Minjakso
COUNTRY : South Korea
GROUP : Product Group

作者：Georgia von le Fort
国家：德国
组别：产业组

AUTHOR : Georgia von le Fort
COUNTRY : Germany
GROUP : Product Group

提交的作品"Ban"旨在通过设计市场和工艺市场的融合创造一个新的市场。它通过对以现代方式重新诠释的传统木制家具的设计进行构图，以各种组合扩展产品，并通过嫁接珍珠母漆等传统工艺技术，将工艺特性添加到产品中。由此，该区域从设计市场扩展到工艺市场。以酒店和美术馆为目标的高端传统工艺产品原型已经完成，使用环保饰面材料的低端产品计划推出。

"Ban", the work submitted, is planned with the goal of creating a new market through the convergence of the design market and the craft market.It is designed to expand the product in various combinations by patterning the design of traditional wooden furniture that has been reinterpreted in a modern way, and the properties of craft are added to the product by grafting traditional craft techniques such as lacquer mother-of-pearl.Via this, the area was expanded from the design market to the craft market. Prototypes for high-end products incorporating high-end traditional crafts have been completed targeting hotels and art galleries, and low-end products using eco-friendly finishing materials are scheduled to be launched.

Relics 是由回收的瓷器制成的容器，可以帮助水果和蔬菜保存更长时间。容器内的托盘可通过蒸发冷却降低温度。这为储存水果和蔬菜创造了最佳条件。本产品可减少浪费，并支持节约资源的生活方式。

Relics are containers made from recycled porcelain that help fruit and vegetables to last longer.A tray inside the container can decrease the temperature through evaporative cooling. This creates optimal conditions for storing fruit and vegetables.With this product, waste is reduced and a resource-saving lifestyle is supported.

"变"几
CHANGEABLE TABLE

作者：李雅云 孙雪松 兰钊 王凤茹 林永贤
机构：中德（泉州）工业设计研究院有限公司
国家：中国
组别：产业组

AUTHOR : Li Yayun, Sun Xuesong, Lan Zhao, Wang Fengru, Lin Yongxian
UNIT : Sino German (Quanzhou) Industrial Design and Research Institute Co., Ltd.
COUNTRY : China
GROUP : Product Group

该款产品聚焦当下年轻人的生活习惯，以生活为内驱力，聚焦生活日用存在的痛点，通过推拉的形式做到形随心变，实现产品造型与功能的高度融合。以茶几和推拉结构结合的突破式创新，为年轻一代的舒适生活助力！

This product focuses on the life habits of young people, takes life as the driving force, focuses on the pain points of daily life, and changes its shape with the heart through pushing and pulling, realizing a high integration of product shape and function. Breakthrough innovation based on the combination of tea table and push-pull structure helps the young generation live a comfortable life!

F=MG

作者：Jakob Glasner
国家：奥地利
组别：产业组

AUTHOR : Jakob Glasner
COUNTRY : Austria
GROUP : Product Group

F=MG 是一款基于无磨损夹紧技术的模块化高架床系统，智能地利用重力，而不是对抗重力。

F=MG is a modular loft bed system based on a wearless clamping technique, using gravity intelligently instead of fighting against it.

心冥想座具便携系列·小忍
LITTLE PATIENCE [MEDITATION SEAT(PORTABLE)]

笑椅
SMILE — CHAIR

作者 : 高凤麟 刘沛桐
机构 : 心冥想健康科技有限公司
国家 : 中国
组别 : 产业组

AUTHOR : Gao Fenglin, Liu Peitong
UNIT : Shine Meditation Health Technology（Hangzhou）Co., Ltd.
COUNTRY : China
GROUP : Product Group

小忍（冥想座具）是一款基于人体工程学的便携家具，它的功能是帮助人们在冥想时轻松完成盘腿坐姿。

Little patience (meditation seat) is a portable furniture based on ergonomics. Its function is to help people easily complete cross-legged sitting posture during meditation.

作者 : 徐乐 包力源 翟伟民 林幸民 辜弯婉
机构 : 杭州大巧创意设计有限公司 / 浙江工业大学之江学院
国家 : 中国
组别 : 产业组

AUTHOR : Xu Le, Bao Liyuan, Zhai Weimin, Lin Xingmin, Gu Wanwan
UNIT : Hangzhou DAQIAO Home Design Studio/Zhijiang College of Zhejiang University of Technology
COUNTRY : China
GROUP : Product Group

笑椅是一把集美学和功能于一体的椅子。椅子的靠背弧度犹如微微上扬的嘴角，故名"笑椅"。 笑椅在工艺上采用先进的冷压曲木技术，用最少的材料来实现天然实木的高贵素材感，也充分运用了木材本身的韧性。严苛细致的制作工艺，传递匠心温暖，让使用者拥有一颗爱物之心。

A chair that combines aesthetics and function.The backrest of the chair is like a slightly raised corner of the mouth, hence the name "Smile— Chair". The chair adopts advanced cold-curved wood technology in the process, uses the least materials to achieve the noble material sense of natural solid wood, and also fully uses the toughness of the wood itself. The rigorous and meticulous production process conveys the warmth of ingenuity and allows the owner to produce a heart of love.

12° Lamp

FITNESS PARCOURS

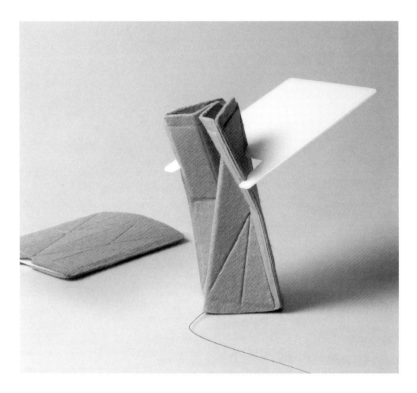

作者 : Kun Geng
机构 : FOM STUDIO SRLS
国家 : 意大利
组别 : 概念组

AUTHOR : Kun Geng
UNIT : FOM STUDIO SRLS
COUNTRY : Italy
GROUP : Concept Group

作者 : Sigi Ramoser, Nico Pritzl
国家 : 奥地利
组别 : 产业组

AUTHOR : Sigi Ramoser, Nico Pritzl
COUNTRY : Austria
GROUP : Product Group

只需几步即可将信封变成一盏灯，12°便携灯给您前所未有的用光体验，它充分利用了 OLED 的特性：环保、超轻薄、光质好、简单。12°手提灯由两个主要部分组成：包装（也是灯杆）和 OLED 面板，两部分在产品的整个生命周期中紧密结合，最大限度地减少资源的使用，增强环保性对 OLED 技术的印象。

An envelope can be changed into a lamp within just several steps, the 12° portable lamp gives you an unprecedented experience of using light, it makes the full use of OLED characteristics: eco-friendly, super light and thin, good quality of light and simple. The 12° portable lamp consists of two main parts: packaging (also the lamppost) and the OLED panel, both parts are closely integrated with each other through the whole lifecycle of the product, which minimizes the usage of resources and strengthens the eco-friendly impression of OLED technology.

多恩比恩市正在扩建其体育设施，在多恩比恩河沿岸修建了一条宽阔的健身步道。这条道路已经敞开了大门，让人们汗流浃背。在美丽的大自然中进行户外锻炼，并配备高质量的设备，现在还配备了一个清晰易懂的指导系统。

The city of Dornbirn is expanding its sports facilities with an extensive fitness trail built along the river of Dornbirn. The course has already opened its doors and invites people to sweat. Outdoor exercises along with high-quality equipment in the middle of a beautiful nature, are now accompanied with a clearly understandable guidance system.

随"变"换
CHANGEABLE LENS ADJUSTABLE PUPIL DISTANCE GLASSES

作者：左叶 张森
机构：温州市森一眼镜设计有限公司
国家：中国
组别：概念组

AUTHOR : Zuo Ye, Zhang Sen
UNIT : Senee Eyewear Design Studio
COUNTRY : China
GROUP : Concept Group

可根据度数变化更换镜片、可调节瞳距的眼镜，解决了年龄增长度数增加变化的问题，满足了不同用户的个性化视光需求。

The lens can be changed according to the power change, and the pupil distance glasses can be adjusted to solve the problem of power increase and change as the age increases, so as to meet the personalized optometry needs of different users.

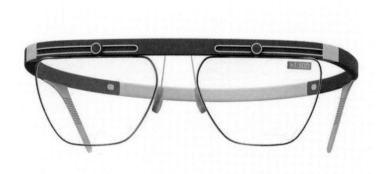

再生混凝土海洋消波块微生态设计
MICRO ECOLOGICAL DESIGN OF RECYCLED CONCRETE OCEAN WAVE ABSORBING BLOCK

作者：龙香华 刘佳豪 谢仁科
国家：中国
组别：概念组

AUTHOR : Long Xianghua, Liu Jiahao, Xie Renke
COUNTRY : China
GROUP : Concept Group

产品采用城市建筑废料制作而成，内部设计有运用潮汐能发电装置，通过潮汐动能带来可持续的绿色能源理念。

The product is made of urban construction waste, and is internally designed with a tidal power generation device, which brings the concept of sustainable green energy through tidal kinetic energy.

厕田
FIELD SANITATION UNIT

BEACH ENVELOPED IN MIST

作者：杨隽 陈向阳
机构：杭州玩家家居设计有限公司
国家：中国
组别：概念组

AUTHOR : Yang Jun, Chen Xiangyang
UNIT : Hangzhou Spieler Design Pty. Ltd.
COUNTRY : China
GROUP : Concept Group

作者：Reiko Kitora, Atsuhito Kitora
国家：日本
组别：概念组

AUTHOR : Reiko Kitora, Atsuhito Kitora
COUNTRY : Japan
GROUP : Concept Group

"厕田"是一款提倡污水就地资源化的公共洗手间。在"双碳"目标之下，"厕田"将成为未来实施碳中和的核心。

"Field Sanitation Unit"is a public toilet that advocates in-situ sewage up-cycling to showcase the possibilities of urban farming and carbon-neutral.

湖底或海底的死海藻或水草引起的臭味、缺氧水体或蓝潮可以通过回收营养盐来解决。通过使用可漂浮块种植海藻和水草，可以节省劳动力并以低成本解决富营养化问题，这些海藻和水藻在生长过程中会吸收营养盐。海藻和水草在生长过程中吸收营养盐并将二氧化碳转化为多糖。该物种拥有丰富的葡萄糖、乙醇和可生物降解塑料的来源，可以促进有效的碳循环。可浮动块吸收并缓慢释放营养盐溶液，使海藻能够在贫营养水域养殖。

Bad odor, hypoxic water mass, or blue tide caused by dead seaweed or waterweed at the bottom of lake or sea can be solved by recovering nutrient salts. Labor-saving and low-cost solution to eutrophication is possible by using floatable blocks for cultivating seaweed and waterweed that take in nutrient salts during growth.Seaweed and waterweed during growth absorb nutrient salt and convert CO_2 to polysaccharide. The species having abundant glucose, a source of ethanol and biodegradable plastic, boosts efficient carbon cycle. Floatable block to absorb and slowly release nutrient salt solution enables seaweed farming in oligotrophic waters.

THINKPAD PLASTIC FREE PACKAGING

作者 : Lenovo Design Innovation Team
机构 : Lenovo
国家 : 中国
组别 : 产业组

AUTHOR : Lenovo Design Innovation Team
UNIT : Lenovo
COUNTRY : China
GROUP : Product Group

ThinkPad Z 系列采用完全无塑料的包装, 非常注重可持续性。

Completely plastic-free packaging designed for the ThinkPad Z series with a strong focus on sustainability.

SUSTAINABLE SHOPPING BAG

作者 : Lim Sungmook
国家 : 韩国
组别 : 产业组

AUTHOR : Lim Sungmook
COUNTRY : South Korea
GROUP : Product Group

这不是用纸制成的购物袋。可多次使用的 "环保购物袋" 坚固耐用, 最多可容纳7千克 (建议4千克) 的物品。这款购物袋甚至还有防水功能, 因此也适合在下雨天或容纳潮湿的物品时使用, 因为它是由100% 可回收材料 Tyvek (HDPE) 制成的。

IT'S NOT PAPER. Multi-usable "Sustainable Shopping Bag" is strong enough to hold up a maximum of 7kg (recommend 4kg). It's even waterproof, so it's good for rain or wet items. Because it's made of 100% recyclable materials Tyvek (HDPE).

常青
EVERGREEN

作者：杜聪 张湘钰 崔喆 米男男 张书炜
机构：合肥联宝信息技术有限公司
国家：中国（台湾）
组别：产业组

AUTHOR : Du Cong, Zhang Xiangyu, Cui Zhe, Mi Nannan, Zhang Shuwei
UNIT : Hefei LCFC Information Technology Co., Ltd.
COUNTRY : Taiwan, China
GROUP : Product Group

自带环保属性，使用 FSC 认证纸张与100% 回收塑料，其中海洋塑料含量从20% 增加到30%，材料性能处于行业领先水平；热塑件具有大理石纹理，用户可以重复利用来种植绿植，延续包装生命周期，减少对环境的影响；用户扫描二维码，上传环保行为照片，增加参与环保的互动性和趣味性，鼓励用户与企业一起参与环保。

It's sustainable, which consists of FSC paper and 100% PRC, and in the cushion OBP increased from 20% to 30%, and the physical property is the best; the cushion has marble texture, which can be reused as flower pots; interactive plan encourage consumers to join in environmental protection.

KENYAN UPCYCLED UNIFORMS

作者：Alex Musembi, Elmar Stroomer, Kirsty Zeller, Khalid Awale
机构：Africa Collect Textiles
国家：肯尼亚
组别：产业组

AUTHOR : Alex Musembi, Elmar Stroomer, Kirsty Zeller, Khalid Awale
UNIT : Africa Collect Textiles
COUNTRY : Kenya
GROUP : Product Group

Africa Collect Textiles（ACT）是一家收集废旧纺织品进行再利用、升级改造和回收的社会企业。ACT 的一项服务为将公司制服回收并制成袋子和背包。我们通过将制服变成最终的品牌和筹款资产，赋予制服新的生命和意义。

Africa Collect Textiles (ACT) is a social enterprise that collects used textiles for reuse, upcycling and recycling. One of ACT's services includes the recycling of corporate uniforms into bags and backpacks. We give new life and meaning to uniforms by turning them into the ultimate branding and fundraising assets.

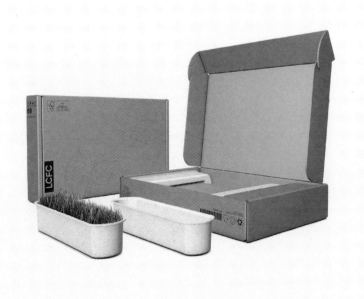

三江源雪域牦牛绒围巾公益礼品
SANJIANGYUAN CHARITY GIFTS

作者：黄莎莉 张安吉 曾令波
机构：腾讯科技（深圳）有限公司 / 深圳慢物质文化创意有限公司
国家：中国
组别：产业组

AUTHOR : Huang Shali, Zhang Anji, Zeng Lingbo
UNIT : Tencent Technology (Shenzhen) Co., Ltd. / Shenzhen Slow Material Culture Creative Co., Ltd.
COUNTRY : China
GROUP : Product Group

通过产品包装设计，呼吁大众关注三江源水源保护，同时帮助当地牧人获得一份收入，带动当地的可持续发展。

Through the product packaging design, the public is called to pay attention to water protection of sanjiangyuan, and at the same time, the local herders are helped to obtain an income and promote the sustainable development of the local area.

循环直运快递箱 —— 为快递包装"绿色化"发展而设计
INFINITE LOOP BOX — "GREEN" FOR EXPRESS PACKAGING

作者：李甫印 江佳婧 陈家超
机构：中荣印刷集团股份有限公司
国家：中国
组别：概念组

AUTHOR : Li Fuyin, Jiang Jiajing, Chen Jiachao
UNIT : Zrp Printing Group Co.,Ltd.
COUNTRY : China
GROUP : Concept Group

这是一款为快递包装"绿色化"发展而设计的产品，循环使用及碳排放量化，构建循环模式及碳交易平台。

This is a design for the "green" development of express packaging, recycling and carbon emission quantification, building a recycling model and a carbon trading platform.

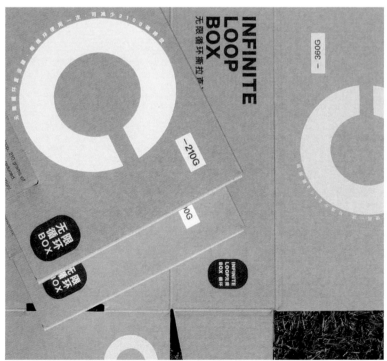

RE-CARDBOARD

"莲花岛号"海洋鱼类培育放生船
"LOTUS ISLAND" MARINE FISH CULTIVATION AND RELEASE SHIP

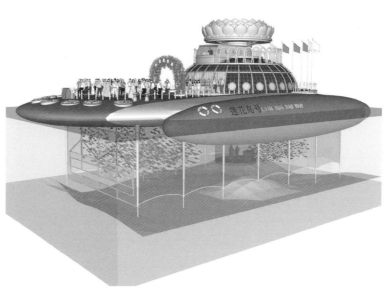

作者：Reiko Kitora, Atsuhito Kitora
国家：日本
组别：概念组

AUTHOR : Reiko Kitora, Atsuhito Kitora
COUNTRY : Japan
GROUP : Concept Group

作者：朱仁民
机构：杭州潘天寿环境艺术设计有限公司
国家：中国
组别：概念组

AUTHOR : Zhu Renmin
UNIT : Hangzhou Pantianshou Environmental Art Design Co., Ltd.
COUNTRY : China
GROUP : Concept Group

救济物资通常被运送到受自然灾害影响的地方。这种纸板包装系统的设计具有二次用途，适用于食物稀缺、每种资源都很重要的地方。盒子里嵌入了装满蔬菜种子的可生物降解塑料胶囊，这样当它们被堆肥时，营养蔬菜就可以从中长出来。废弃的纸板被再生为植物，以抵消二氧化碳排放，同时为需要的人提供食物。Re-cardboard 体现了我们的目标，即帮助人们认识到环境问题是一个紧迫的问题，尽管环境问题往往被认为是全球性的。

产品以"科学培育、慈善放养"的实践，成为拯救海洋鱼类的培育、放生、教育、文旅综合性场所。

Relief supplies are commonly delivered to locations affected by natural disasters. This system of cardboard packaging is designed with a secondary use that is tailored to sites where food is scarce and every resource counts. Biodegradable plastic capsules filled with vegetable seeds are embedded into the boxes so when they are composted, nourishing vegetables can grow out of them. The discarded cardboard is reborn as plants to counteract CO_2 emissions, while simultaneously providing food for people who need it. Re-Cardboard embodies our goal to help people to recognize the environmental problem as an immediate one, although the environmental problem tends to be thought of on a global scale.

This product is a comprehensive site of fish cultivation, release, education and cultural tourism that aims at saving fish in the sea through the practice of "scientific cultivation and charitable stocking".

NIGGBUS

作者 : Sigi Ramoser
国家 : 奥地利
组别 : 产业组

AUTHOR : Sigi Ramoser
COUNTRY : Austria
GROUP : Product Group

为一个重新自然化项目进行沟通。

Comunication for a renaturalisation project.

WIND POWER NEXT-GENERATION

作者 : Reiko Kitora, Atsuhito Kitora
国家 : 日本
组别 : 概念组

AUTHOR : Reiko Kitora, Atsuhito Kitora
COUNTRY : Japan
GROUP : Concept Group

下一代风力发电机是一种利用风力发电的发电机, 但它没有叶片。

Wind power next-generation is a power generator using wind power, but it has no blade.

盲人之诗 —— 可触化共识开源控件
POETRY OF THE VISUAL IMPAIRMENTS

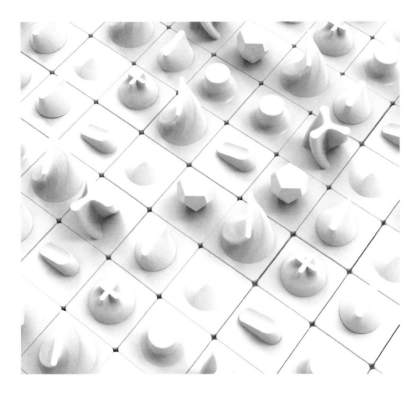

作者：王奕澄 李珊珊
国家：中国
组别：概念组

AUTHOR : Wang Yicheng, Li Shanshan
COUNTRY : China
GROUP : Concept Group

"盲人之诗"是在思辨设计概念下，用触觉反馈辅助盲人自理，进而启发更广泛的对生命尊严探讨的社会创新设计。

"Poetry of the visual impairments "is a social innovation design works, which under the concept of speculative design that uses tactile feedback to assist the visual impairments to take care of themselves, thus inspiring a broader discussion on the dignity of life.

生活智慧
LIFE WISDOM

ALPHAMOD

光合星球:儿童植物听诊器
PLANNNET: PLANT STETHOSCOPE

作者：Danilo Carmelo De Paola
国家：意大利
组别：产业组

AUTHOR : Danilo Carmelo De Paola
COUNTRY : Italy
GROUP : Product Group

作者：王昊远 陈文丽 汤宇萱 杨馥嫚
国家：中国
组别：概念组

AUTHOR : Wang Haoyuan, Chen Wenli, Tang Yuxuan, Yang Fuman
COUNTRY : China
GROUP : Concept Group

AlphaMod 是一款玩具，其目的是通过字母和字符学习外语，教育学龄儿童融入社会；所有这些都是通过一种有趣的，同时又是社会化的学习方法来实现的。

AlphaMod is a toy, the aim of which is to educate school-age children, about inclusion through the learning of foreign languages through alphabets and characters; all through a fun and at the same time socializing method of learning.

PLANNNET 是一款能将植物生物电实时转译成乐符的交互玩具，引导儿童聆听植物的声音，从而更好地理解自然与生命的意义。

An interactive toy named PLANNNET, which can translate plant's bioelectrical signals into MIDI phrases, is designed to help children listen to the language of plants and cultivate their love for nature and life.

速绘多功能键盘 DK001
DRAWING KEYBOARD DK001

华章
TOTEBOOK

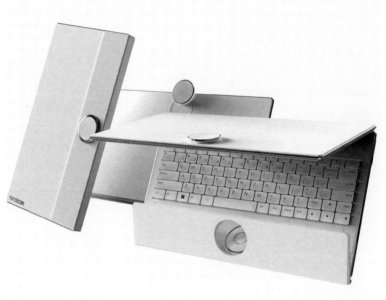

作者：莫倩 杨灿烈
机构：深圳市绘王动漫科技有限公司
国家：中国
组别：产业组

AUTHOR : Mo Qian, Yang Canlie
UNIT : Shenzhen Huion Animation Technology Co.,Ltd.
COUNTRY : China
GROUP : Product Group

作者：田乔 刘德智 王文东 黄孟凯 袁康
机构：合肥联宝信息技术有限公司
国家：中国
组别：产业组

AUTHOR : Tian Qiao, Liu Dezhi, Wang Wendong, Huang Mengkai, Yuan Kang
UNIT : Hefei LCFC Information Technology Co., Ltd.
COUNTRY : China
GROUP : Product Group

DK001是一款可快捷操作的多功能绘图设备，适用众多设计领域，解决用户使用传统键盘绘画设计时操作不便的问题。

DK001 is a multi-function drawing device that can be operated quickly. It is suitable for many design fields and solves the problem of inconvenient operation when users use traditional keyboards to draw and design.

一款适合目前办公生活场景之间高度融合的笔记本电脑。

A laptop suitable for current close integration relationship between office and life scenarios.

联想 Go 无线拆分键盘、联想 Go 无线垂直鼠标
LENOVO GO WIRELESS SPLIT KEYBOARD & VERTICAL MOUSE

MANTA · 一体式足部键盘
MANTA-ONE-PIECE FOOT KEYBOARD

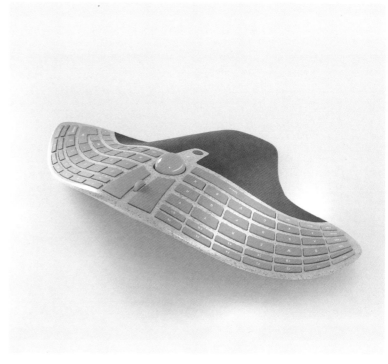

作者：Lenovo Design Innovation Team
机构：联想（北京）有限公司
国家：中国
组别：产业组

AUTHOR : Lenovo Design Innovation Team
UNIT : Lenovo
COUNTRY : China
GROUP : Product Group

作者：王嘉豪 胡宁博 张勇
国家：意大利
组别：概念组

AUTHOR : Wang Jiahao, Hu Ningbo, Zhang Yong
COUNTRY : Italy
GROUP : Concept Group

我们的联想 Go 人体工学套装是为最大限度地满足终端用户的舒适度而设计的，并鼓励最佳姿势来提高生产力。

Our Lenovo Go Ergonomic Set, however, is engineered for maximum end-user comfort and encourages optimal posture to enhance productivity.

以蝠鲼为灵感，通过人体工学设计的方式，解决上肢残障人士操作计算机难题的一体式足部键盘。

Inspired by manta, an integrated foot input device that solves the problem of upper limb disabled people operating computers through ergonomic design.

HAZAR

视障围棋
AOTU GO

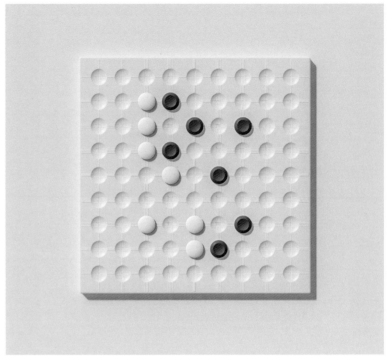

作者：Ozan Tığlıoğlu, Wakako Esra Aras, Furkan Öz
机构：GOA Design Factory
国家：土耳其
组别：产业组

AUTHOR : Ozan Tığlıoğlu, Wakako Esra Aras, Furkan Öz
UNIT : GOA Design Factory
COUNTRY : Turkey
GROUP : Product Group

可移动和模块化的办公桌设计，带有隔板和照明，适用于办公室和集体工作区 。

Movable and modular desk design with separator and lighting for offices and collective work areas.

作者：高晟晏 赖孝文 王炯捷 周超
机构：电子科技大学中山学院
国家：中国
组别：概念组

AUTHOR : Gao Shengyan, Lai Xiaowen, Wang Jiongjie, Zhou Chao
UNIT : Zhongshan Institute,University of Electronic Science and Technology
COUNTRY : China
GROUP : Concept Group

一款创新的围棋套装，为视障人士提供娱乐，且能与视力正常者下棋。

A creative Go suit designed for visually impaired individuals , can afford them entertainment and chess with visually-normal people.

ADDFUN AFX 自由屏
ADDFUN AFX

作者：万虹亦 张焕峰 宋丽娴
机构：四川长虹电子控股集团有限公司
国家：中国
组别：产业组

AUTHOR : Wan Hongyi, Zhang Huanfeng, Song Lixian
UNIT : Sichuan Changhong Electronics Holding Group., Ltd.
COUNTRY : China
GROUP : Product Group

AFX 是一款以用户使用体验为中心的设计产品。它是一台多功能娱乐机，集学习、观影、娱乐为一体。

AFX is a design product centered on the user experience. It is a multi-functional entertainment machine that integrates learning, watching movies and entertainment.

智能充气游泳圈
INTELLIGENT INFLATABLE SWIMMING RING

作者：代明汲 陈浩
机构：义乌乐樱电子商务有限公司 / 杭州云丘工业设计有限公司
国家：中国
组别：产业组

AUTHOR : Dai Mingji, Chen Hao
UNIT : Yiwu Leying E-Commerce Co., Ltd. / Hangzhou Yunqiu Industrial Design Co., Ltd.
COUNTRY : China
GROUP : Product Group

以创新性的充气方式、结构形式和功能，提升产品的用户体验、行业产品的专业度，高效利用社会资源。

With innovative inflatable methods, structural forms and functions, we can improve the user experience of products, the professionalism of industry products, and make efficient use of social resources.

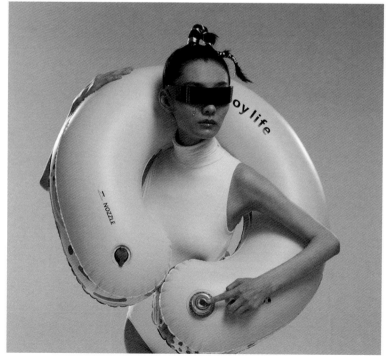

DATA HUMIDIFIER

MOMENTUM — A CHAIR THAT WANTED TO LIVE

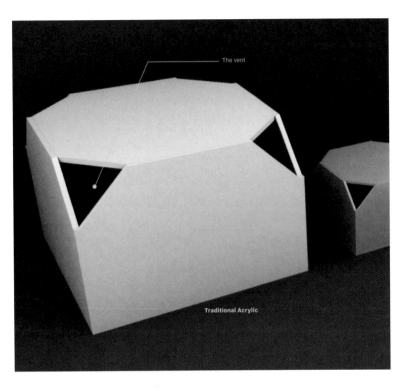

作者：Yao Yao
国家：美国
组别：概念组

AUTHOR：Yao Yao
COUNTRY：United States
GROUP：Concept Group

作者：Joanna Rudzińska
国家：波兰
组别：概念组

AUTHOR：Joanna Rudzińska
COUNTRY：Poland
GROUP：Concept Group

随着技术和大数据时代的到来，个人隐私和信息泄露越来越严重，整个过程不透明。通过研究，设计者发现了隐私可控性的重要性。许多应用程序模糊了用户的知情权，这无疑会使用户的信息在不知情的情况下泄露，并使用户体验更加糟糕。隐私变成了一个大谎言。数据加湿器使用不同的气味来通知用户在使用互联网时存在隐私安全问题，如秘密数据收集。使用这款智能加湿器，用户能够根据发出的相应气味获得隐私使用通知。

With the advent of technology and big data, personal privacy and information leakage are becoming increasingly serious, and the entire process is opaque. Through research, designers have discovered the importance of privacy controllability. Many applications blur users' right to know, which undoubtedly leads to the leakage of information without their knowledge and worsens the user experience. Privacy has become a big lie.Data humidifiers use different odors to notify users of privacy and security issues when using the internet, such as secret data collection. With this intelligent humidifier, users can receive privacy usage notifications based on the corresponding odor emitted.

Momentum 是一款模块化家具，旨在解决人们在办公室久坐不动的问题。这款家具的形状有助于提供更多的身体姿势可能性，这可以通过扭转其模块来扩展。Momentum 通过引入刺激 - 振动来支持姿势的变化，以指示新身体姿势的时间。

Momentum is a modular piece of furniture addressing the problem of sedentary behavior in the office. The shape of the furniture facilitates more body position opportunities, which can be expanded by twisting its modules. Momentum supports position change by introducing stimulus - vibration indicating time for a new body posture.

SOFI

作者：Rory Aartsen
国家：荷兰
组别：概念组

AUTHOR : Rory Aartsen
COUNTRY : Dutch
GROUP : Concept Group

Sofi 是一种设备，旨在通过触发对同事的共同责任来促进工作场所的重要休息，从而提高办公室工作人员的幸福感和工作效率。Sofi 通过调整琴弦的张力，将集体和个人的活动水平转换成一个音调高度。

Sofi is a device meant to prompt vital breaks in the workspace by triggering co-responsibility towards co-workers improving the well-being and productivity of office workers. Sofi translates both the collective and individual activity levels into a tone height by adjusting the tension of the strings.

BOE 95 8K 智慧终端
BOE 95 8K SMART TERMINAL

作者：索睿睿 王艺 赵玉盼 王子锋 李照威
机构：京东方科技集团股份有限公司
国家：中国
组别：产业组

AUTHOR : Suo Ruirui, Wang Yi, Zhao Yupan, Wang Zifeng, Li Zhaowei
UNIT : BOE Technology Group Co., Ltd.
COUNTRY : China
GROUP : Product Group

BOE 95 8K 智慧终端是一款结合屏幕模组分离和磁吸式组装设计达到 6mm 超薄厚度的高端商用显示设备。

BOE 95 8K Smart Terminal is a high-end commercial display device that combines screen module separation and magnetic assembly design to achieve a 6mm ultra-thin thickness.

LOMARE MEMORY — SAVE EMISSIONS BY LOWERING POWER!

UNPLUQ TAG

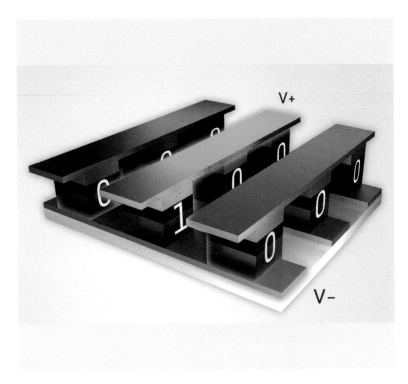

作者：Evgeniy Donchev, Bin Zou, Andrei Mihai, Jan Zemen, Jisheng Sun
机构：LoMaRe Technologies Limited
国家：英国
组别：概念组

AUTHOR : Evgeniy Donchev, Bin Zou, Andrei Mihai, Jan Zemen, Jisheng Sun
UNIT : LoMaRe Technologies Limited
COUNTRY : UK
GROUP : Concept Group

作者：Jorn Rigter, Tim Smits
机构：Unpluq Technology B.V
国家：荷兰
组别：产业组

AUTHOR : Jorn Rigter, Tim Smits
UNIT : Unpluq Technology B.V
COUNTRY : Dutch
GROUP : Product Group

LoMaRe Memory 是一种全新的非易失性 RAM 技术，通过采用全新的材料组合解决了存储和处理数据时的高功率需求问题，这是以前从未用于存储器应用的技术，也是科学的一次范式转变，将成为未来技术的发展方向，就在今天！

LoMaRe Memory is a new non-volatile RAM technology solving the problem of high power requirements when storing and processing data by utilising new material combinations, never used before for memory applications, and a paradigm shift in the science that results in a future technology, today!

Unpluq 标签是世界上第一个物理干扰屏障，可帮助您阻止让您分散注意力的应用程序并提高工作效率。学会有意识地使用手机，重新掌控您自己的时间和注意力。每天可节省1小时以上的屏幕使用时间。

The Unpluq Tag is the world's first physical distraction barrier that helps You Block Distracting Apps & Be More Productive. Learn to use your phone mindfully & regain control of your time and attention. Save more than 1 hour of screen time every day.

LINX

ROTATA

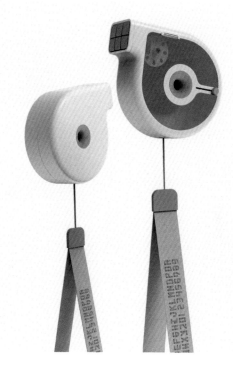

作者：Saskia Buana
国家：中国（台湾）
组别：产业组

AUTHOR : Saskia Buana
COUNTRY : Taiwan, China
GROUP : Product Group

作者：王小举 原炳坤 孙矩韬 杨周 陈淑怡
机构：华东理工大学 / 景德镇陶瓷大学
国家：韩国
组别：概念组

AUTHOR : Wang Xiaoju, Yuan Bingkun, Sun Jutao, Yang Zhou, Cheng Shuyi
UNIT : East Chinese University of Science and Technology / Jingdezhen Ceramic Institute
COUNTRY : South Korea
GROUP : Concept Group

Linx 是一款面向休闲视频博主的模块化拍摄套件，旨在简化内容创作者的拍摄过程，并让他们能够更好地控制拍摄过程，尤其是在独自拍摄时 。

Linx is a modular filming kit targeted for the casual vlogger, aimed to simplify and give content creators more control of their filming process, especially when filming alone.

ROTATA 盲文便签打印机是通过打印盲文可粘贴便签的方式来解决盲人部分生活信息识别需求的产品 。

ROTATA Braille Memo Printer is a product that can print pastable Braille memos to meet the needs of blind people to identify part of their life information.

SQUARE OFF SWAP

作者： Bhavya Gohil, Atur Mehta, Dhiraj Gehlot, Siddhesh Sawant, Umang Shah
机构： Infivention Technologies Pvt. Ltd. / Oro Innovations Pvt. Ltd.
国家： 印度
组别： 产业组

AUTHOR : Bhavya Gohil, Atur Mehta, Dhiraj Gehlot, Siddhesh Sawant, Umang Shah
UNIT : Infivention Technologies Pvt. Ltd. / Oro Innovations Pvt. Ltd.
COUNTRY : India
GROUP : Product Group

Square Off Swap Smart 是一款自动棋盘游戏，可让用户在一个平面上玩多款游戏，包括国际象棋、跳棋、Halma 和 Connect 4。这款下一代棋盘用途广泛，具有超快速闪电战动作和交互式嵌入式灯光，凭借其物理体验，为游戏之夜作好准备。

Square off Swap smart & an automated board game that lets users play multiple games including Chess, Checkers, Halma and Connect 4 on a single surface. This next-gen board is versatile with super fast blitz movements and interactive embedded lights making it game nights ready with its physical experience.

未知实验室001号吉他
U-LAB 001 GUITAR

作者： 陈锋明 林思婷 陈伟浩 王思源 曾嘉荣
机构： 未知星系科技（深圳）有限公司 / 深圳市格外设计经营有限公司
国家： 中国
组别： 产业组

AUTHOR : Chen Fengming, Lin Siting, Chen Weihao, Wang Siyuan, Zeng Jiarong
UNIT : Unknown Galaxy Technology (Shenzhen) Limited / inDare Design Strategy Limited
COUNTRY : China
GROUP : Product Group

U-LAB 001 Guitar 是一款专为音乐爱好者设计的智能吉他，让更多的乐迷享受演奏音乐的乐趣，不必为学习吉他技法而烦恼。

U-LAB 001 Guitar is a smart guitar designed for music lovers. Let more music fans enjoy the fun of playing music without having to worry about learning guitar skills.

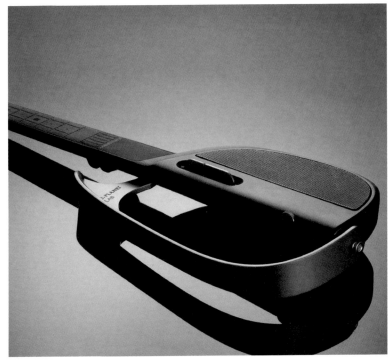

SUPERSPACE

作者 : Jarn Bulling, Will Grant, John Ditchburn, Ant Erwin, Craig Spencer
机构 : Everplay Ltd., 4DESIGN
国家 : 新西兰
组别 : 产业组

AUTHOR : Jarn Bulling, Will Grant, John Ditchburn, Ant Erwin, Craig Spencer
UNIT : Everplay Ltd., 4DESIGN
COUNTRY : New Zealand
GROUP : Product Group

SUPERSPACE: 一个为儿童设计的磁性模块化游戏空间构建器。真人大小的磁性面板让孩子们能够创造出几乎任何他们能够想象到的结构，并在他们自己的创作中玩耍。

SUPERSPACE: A magnetic modular play space builder for kids. Life-size magnetic panels allow children to create almost any structure they can dream up and play in their own creation.

飞莫水光针二代
Himeso II

作者 : 杨洋
机构 : 宁波菲莫智能科技有限公司 / 宁波可点工业设计有限公司
国家 : 中国
组别 : 产业组

AUTHOR : Yang Yang
UNIT : Femooi / Ningbo COOR Industrial Design Co., Ltd.
COUNTRY : China
GROUP : Product Group

飞莫水光针二代是一款通过智能科技解决女性肌肤问题的便携式家用美容产品。

Himeso II is a portable household beauty instrument that solves women's skin problems through intelligent technology.

THE EYE-SEEDS

作者 : Atsuhito Kitora, Reiko Kitora
国家 : 日本
组别 : 概念组

AUTHOR : Atsuhito Kitora, Reiko Kitora
COUNTRY : Japan
GROUP : Concept Group

EYE-SEEDS 是一款特殊的玻璃材质（脂肪酸聚酯）， 这种材质可以保护您的眼睛免受细小灰尘的侵害。 产品结构简单，轻量且柔韧，不易碎。 使用完毕后会自发降解为水和二氧化碳，并激活玻璃中休眠的植物种子。这些养分被植物重新吸收循环。

EYE-SEEDS is a unique glass material (fatty acid polyester) designed to protect your eyes from fine dust particles. The product features a simple, lightweight, and flexible structure that is resistant to breakage. After use, it spontaneously degrades into water and carbon dioxide, activating dormant plant seeds contained within the glass. These nutrients are then absorbed and recycled by the plants.

ONETWO

作者 : Umut Demirel, Elif Ergur, Safa Akyuz
国家 : 土耳其
组别 : 概念组

AUTHOR : Umut Demirel, Elif Ergur, Safa Akyuz
COUNTRY : Turkey
GROUP : Concept Group

Onetow 是一种包装概念，将牙膏和牙线放在同一个容器中，以鼓励人们每天使用牙线。

Onetwo is a packaging concept that brings toothpaste and floss together in one single container to encourage people to floss every day.

箭牌 VISWASH 智能坐便器
ARROW VISWASH SMART TOILET

作者：刘志强 封涵宇 朱重华 黄河 刘芬
机构：广东乐华智能卫浴有限公司 / 广州美术学院
国家：中国
组别：产业组

AUTHOR : Liu Zhiqiang, Feng Hanyu, Zhu Chonghua, Huang He, Liu Fen
UNIT : Guangdong Lehua Intelligent Sanitary Ware Co., Ltd. / Guangzhou Art College
COUNTRY : China
GROUP : Product Group

VISWASH 是一款面向高端商场的可视化自洁座圈坐便器，通过可视化自清洁座圈的方式减少用户对公共马桶的膈应心理。

VISWASH is a visual self-cleaning seat toilet for high-end shopping malls, which reduces users' frustration with public toilets by visualizing the self-cleaning seat.

VESTEL T40 洗衣机
T40 WASHING MACHINE

作者：M.Saner Öztürkler, Vestel White Industrial Design Team
机构：Vestel Beyaz Eşya San. ve Tic.AŞ.
国家：土耳其
组别：产业组

AUTHOR : M.Saner Öztürkler, Vestel White Industrial Design Team
UNIT : Vestel Beyaz Eşya San. ve Tic.AŞ.
COUNTRY : Turkey
GROUP : Product Group

Vestel T40洗衣机的设计目的是为用户在日常生活中遇到的问题提供解决方案。除了蒸汽发生器技术外，还有许多环保创新。新开发的 Sterilizone ® 消毒技术能够对各种细菌和病毒进行消毒。

Vestel T40 is designed to bring solutions to problems that users are facing in their daily life. Besides the steam generator technology, there are many environmentally friendly innovations. The newly developed Sterilizone ® Disinfection Technology enables the disinfection of a wide variety of bacteria and viruses.

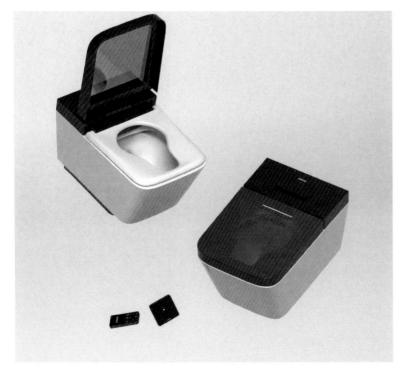

AWAMIST

作者：Shinya Mizutani, Chiaki Murata, Motoyoshi Ono, Yoshitaro Yamanaka, Makoto Ise
机构：MIZUTANI VALVE M.F.G. Co., Ltd. / Hers Design Inc.
国家：日本
组别：产业组

AUTHOR : Shinya Mizutani, Chiaki Murata, Motoyoshi Ono, Yoshitaro Yamanaka, Makoto Ise
UNIT : MIZUTANI VALVE M.F.G. Co., Ltd. / Hers Design Inc.
COUNTRY : Japan
GROUP : Product Group

这是一种非接触式自动洗手装置，设计用于安装在洗手间以外的地方，如玄关、客厅等。独创的水谷自动阀门系统提供了高清洁力、节水、卫生的洗手方式，并防止水溅出。

This is a non-contact automatic handwashing device designed to be installed in places other than washrooms, such as entryways and living rooms. The originally developed Mizutani Auto Valve System provides high cleaning power, water saving, hygienic hand washing, and prevention of water splashing.

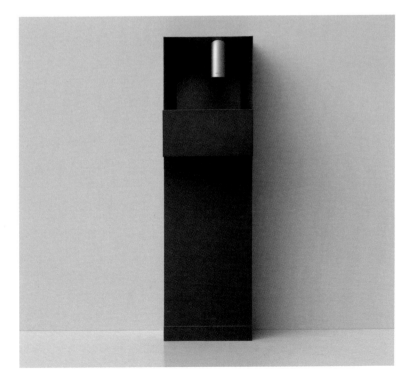

TOOTH.ECO

作者：Joshua Oates, Kiana Guyon, Oliver Grime
机构：Eco Tooth Ltd.
国家：英国
组别：产业组

AUTHOR : Joshua Oates, Kiana Guyon, Oliver Grime
UNIT : Eco Tooth Ltd.
COUNTRY : UK
GROUP : Product Group

你扔掉的每一把牙刷都依然存在在这个世界上。全世界每年生产超过47亿支牙刷，其中99%以上是不可生物降解的。解决方案是可持续设计的口腔护理产品。

Every single toothbrush you have ever thrown away is still in existence. Over 4.7 Billion toothbrushes are produced worldwide every-year and over 99% of these are not biodegradable. The solution is sustainably designed oral care.

箭牌 NOVASH 智能坐便器
ARROW NOVASH SMART TOILET

作者：刘志强 李宇宽 朱重华 黄河
机构：广州美术学院
国家：中国
组别：产业组

AUTHOR : Liu Zhiqiang, Li Yukuan, Zhu Zhonghua, Huang He
UNIT : The Guangzhou Academy of Fine Arts
COUNTRY : China
GROUP : Product Group

Novash 坐便器是一款通过一体化自洁马桶座圈的方式消除用户对公共卫生间坐便器心理隔阂的智能坐便器产品。

Novash Toilet is a smart toilet product that eliminates the psychological barrier of users to public bathroom toilets by means of an integrated self-cleaning toilet seat.

倍至便携冲牙器 A30
BIXDO ULTRA COMPACT WATER FLOSSER (A30)

作者：谭丞深 韩硕
机构：上海飞象健康科技有限公司
国家：中国
组别：产业组

AUTHOR : Tan Chengshen, Han Shuo
UNIT : Bixdo(SH) Healthcare Technology Co.,Ltd.
COUNTRY : China
GROUP : Product Group

A30是一款便携式胶囊洗牙器，专为满足人们外出随时洁牙的需求而设计。设计团队以功能性、实用性和便携性为核心理念对洗牙器的结构进行了优化，将储水箱以可伸缩的形式融入机身中，以0.64mm聚焦水线可实现比牙刷、牙线更深入牙龈沟和齿缝的清洁，协助牙齿正畸，杜绝龋齿、牙周炎等口腔问题的发生。

A30 is a portable water flosser designed to meet the need for on-the-go oral hygiene. The design team focused on functionality, practicality, and portability, optimizing the structure of the water flosser. They integrated the water reservoir in a collapsible form into the body of the device, and with a 0.64mm focused water stream, it can achieve a deeper cleaning in gum pockets and between teeth compared to toothbrushes and dental floss. This helps with teeth alignment and prevents oral issues like cavities and periodontal disease.

ALBERT — IMPROVING PARKINSON'S MOBILITY

作者 : Baptiste Maingon, Antoine Beynel
机构 : Miio Studio
国家 : 中国
组别 : 概念组

AUTHOR : Baptiste Maingon, Antoine Beynel
UNIT : Miio Studio
COUNTRY : China
GROUP : Concept Group

我们为老年人开发了一款智能手杖（主要针对患有帕金森神经退行性疾病的老年人）。我们的产品可以帮助他们自己和他们的亲人监测疾病进展，同时改善患者的日常生活。

We developed a smart walking stick for elderly people (mostly focusing on elderly with Parkinson's neuro-degenerative disease), Our product help them and their beloved ones to monitor the disease evolution while improving the patient's daily life.

LIGHT GLASS

作者 : Zun Wang
机构 : Eindhoven University of Technology
国家 : 荷兰
组别 : 概念组

AUTHOR : Zun Wang
UNIT : Eindhoven University of Technology
COUNTRY : Dutch
GROUP : Concept Group

LightGlass 是一款针对康复期心脏病患者的运动计时推荐系统。它利用灯光的变化作为一种抽象的数据表示，激励用户保持锻炼习惯，以提高他们的健身水平。

LightGlass is an exercise timing recommender system for cardiac patients in the rehabilitation period. It uses the changes of light as an abstract data representation to motivate users to maintain exercise habits to enhance their fitness level.

SCRY 一体增材制造鞋履
SCRY SHUTTLE

作者：魏子雄 程书馨
机构：SCRY Lab
国家：中国
组别：产业组

AUTHOR：Wei Zixiong, Cheng Shuxin
UNIT：SCRY Lab
COUNTRY：China
GROUP：Product Group

SCRY Shuttle 是全球首款"一体增材制造鞋履"，革命性地开创鞋履行业数字化设计制造的新型模式以及新类别一体3D打印鞋。

SCRY Shuttle is the world's first pair of Integrated Additive Manufacturing shoes, revolutionizing the new mode of digital design and manufacturing in the footwear industry and creating a new category of shoes — Integrated 3D printing footwear.

HARNESSWARMER

作者：Stanley KWOK
机构：KnitWarm Limited
国家：中国（香港）
组别：产业组

AUTHOR：Stanley KWOK
UNIT：KnitWarm Limited
COUNTRY：Hong Kong SAR, China
GROUP：Product Group

HarnessWarmer 让您能够以一种用户友好和可持续的方式兼顾温暖与时尚，以解决穿着过多层衣服可能会限制您的活动并增加摔倒的风险的问题，同时老年人更脆弱，更容易摔倒，特别是如果他们还有长期健康问题的话。

HarnessWarmer lets you stay warm and sleek in a user-friendly and sustainable way for solving wearing too many layers of clothing can restrict your movements and increase the risk of falling, but older people are more vulnerable and likely to fall, especially if they have a long-term health condition.

虤 (Yan)
YAN

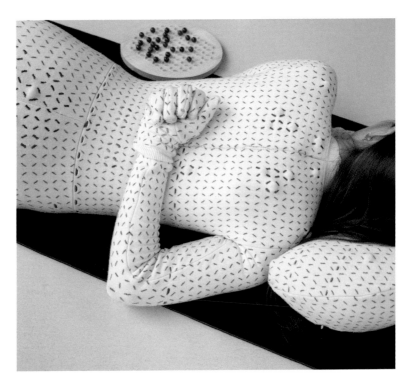

作者：时晓曦 田佳艺 傅海苗
机构：图乐埃（北京）文化有限公司
国家：中国
组别：产业组

AUTHOR : Shi Xiaoxi, Tian Jiayi, Fu Haimiao
UNIT : 2-LA design
COUNTRY : China
GROUP : Product Group

虤 (Yán) 是一款通过 AI 算法与3D 打印技术结合的方式来解决鞋类生产复杂问题的一体鞋产品。

Yan is an integrated shoe that solves the complex problems of footwear production through the combination of AI algorithm and 3D printing technology.

THE HEALING IMPRINT

作者：Laura Deschl
国家：德国
组别：概念组

AUTHOR : Laura Deschl
COUNTRY : Germany
GROUP : Concept Group

Healing Imprint 是一种治疗性服装，外观类似运动服，旨在帮助治愈创伤。通过将指压按摩与基于瑜伽的体验性运动结合，这款服装旨在提高个体的健康和福祉。

The Healing Imprint is a therapeutic garment that looks like activewear but is made to help heal trauma. By combining acupressure with an embodied movement practice based on yoga, the garment seeks to improve an individual's well-being.

COOL

IN TOUCH —— 远程触觉交互衣
IN TOUCH — REMOTE TACTILE INTERACTION SUIT

作者：Andrea Senatori
国家：意大利
组别：产业组

AUTHOR : Andrea Senatori
COUNTRY : Italy
GROUP : Product Group

作者：王菲 林嘉雯 陈超 张书缘 郑晨辉
机构：中国美术学院
国家：中国
组别：概念组

AUTHOR : Wang Fei, Lin Jiawen, Chen Chao, Zhang Shuyuan, Zheng Chenhui
UNIT : China Academy of Art
COUNTRY : China
GROUP : Concept Group

COOL 是一款智能手表。它是一款创新产品，旨在通过将用户的生物信息转化为直观易懂的图形，来测量时间的不同维度，包括年龄时间和用户的私人、个人、情感和心理时间。

COOL is a smartwatch, an innovative product that seeks to measure chronological time as well as the user's private, personal, emotional, and psychological time by transforming the person's biometrics into direct, understandable, and highly intuitive graphics.

In Touch 是一款通过动作捕捉技术识别动态指令进行远程指令传输，并通过振子实现远程触觉传递的衣服。

In Touch is a suit that delivers remote instructions by motion capture technology which recognizes motion instructions, and transmits remote tactile sense by oscillator.

ÉRA PLANET 星球演变"百科全书"—— 结合 AR 的 模块化智能木玩
ÉRA PLANET— MODULAR SMART WOODEN TOY WITH AR

3D 打印学步鞋
3D PRINTED TODDLER SHOES

作者：杨晟 余艾琳 王枫淯 林伟嘉 刘一涵
机构：浙江大学
国家：中国
组别：概念组

AUTHOR : Yang Sheng, Yu Ailin, Wang Fengyu, Lin Weijia, Liu Yihan
UNIT : Zhejiang University
COUNTRY : China
GROUP : Concept Group

作者：许方雷 鲁琪 贾小卜 杨勇 李亚琦
机构：北京斯克莱特科技有限公司
国家：中国
组别：产业组

AUTHOR : Xu Fanglei, Lu Qi, Jia Xiaobu, Yang Yong, Li Yaqi
UNIT : Beijing Scrat 3D Technology Co., Ltd.
COUNTRY : China
GROUP : Product Group

ÉRA-Planet 是一款模块化智能木质玩具，通过 AR，趣味科普地球发展历程，为孩子建立生态观，激发孩子的想象力和创造力。

éra-Planet is a modular smart wooden toy that, through AR, provides a fun way to teach children about the development of the Earth, build an ecological perspective and stimulate their imagination and creativity.

3D 打印学步鞋是一款采用超高速3D 打印技术、自研材料实现专业性能和舒适体验的童鞋，助力宝宝健康学步。

Using ultra-fast 3D printing speed technology, IR&D printing materials have the characteristics of breath ability, cushioning, easy bending, etc., and can be recycled. Perfect for early and beginner walkers.

ON2COOK: WORLD'S FASTEST COOKING DEVICE

魔方灶
MAGIC STOVE

作者：Sanandan Sudhir
机构：InventIndia Innovations Pvt. Ltd.
国家：印度
组别：产业组

AUTHOR : Sanandan Sudhir
UNIT : InventIndia Innovations Pvt. Ltd.
COUNTRY : India
GROUP : Product Group

作者：秦龙 李枢 邵海忠 徐燕 杨紫荆
机构：宁波方太厨具有限公司
国家：中国
组别：产业组

AUTHOR : Qin Long, Li Shu, Shao Haizhong, Xu Yan, Yang Zijing
UNIT : Ningbo Fotile Kitchen Ware Co., Ltd.
COUNTRY : China
GROUP : Product Group

"On2Cook，世界上最快的专利 AI 智能烹饪设备，采用了革命性的组合技术：微波盖 + 火焰 / 电磁炉底座，节省高达70% 的烹饪时间，保留熟食的颜色、质地，营养高达40%；同时附带的应用程序（订购、库存、食谱、分享）可以实现一个健康和能源节约意识的智能社区愿景！"

On2Cook, World's FASTEST PATENTED AI-enabled COOKING DEVICE, uses Revolutionary Combination Technology: Microwave lid + flame/Induction base saving upto 70% cooking time and upto 40% energy retaining colour, texture & nutrients of cooked food; whereas the Accompanying App (Order, Inventory, Recipe, Share). Envisioning a Healthy & Energy Conscious Smart-Community!

魔方灶是一款通过智能折叠、隔空操控等技术，向年轻人传达乐享烹饪、乐享生活理念的创意厨房产品。

The Magic Stove is a creative kitchen product that conveys the concept of enjoying cooking and enjoying life to young people through technologies such as intelligent folding and remote control.

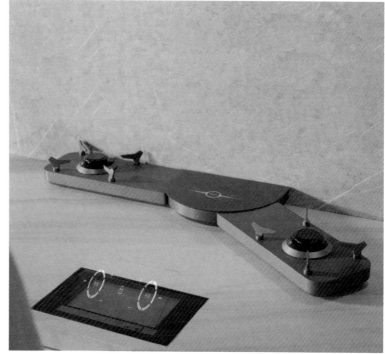

新鲜 —— 智能冰箱助手
FRESH — INTELLIGENT REFRIGERATOR ASSISTANT

MOON DETOX WATER BOTTLE

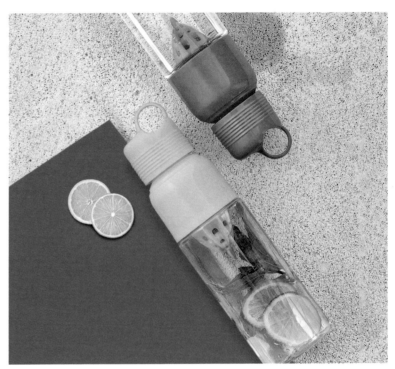

作者：张德寅 沈诚仪 钟方旭 郭和睿 叶琦钧
机构：浙江大学
国家：中国
组别：概念组

AUTHOR : Zhang Deyin, Shen Chengyi, Zhong Fangxu, Guo Herui, Ye Qijun
UNIT : Zhejiang University
COUNTRY : China
GROUP : Concept Group

作者：Fatma Bagcivan, Goksu Oral, Kubra Abinikman, Sena Engin
机构：Mercanlar Kitchenware Inc.
国家：土耳其
组别：产业组

AUTHOR : Fatma Bagcivan, Goksu Oral, Kubra Abinikman, Sena Engin
UNIT : Mercanlar Kitchenware Inc.
COUNTRY : Turkey
GROUP : Product Group

FRESH 是一个解决冰箱食物过度囤积而造成巨大食物浪费问题的智能冰箱助手。

FRESH is a smart refrigerator assistant to solve the problem of huge food waste caused by excessive food storage in the refrigerator.

这款水瓶旨在满足注重健康生活但饮水困难人群的日常排毒需求，目的是通过将水变甜来使人养成饮水习惯。它是由各种可回收的无毒材料制成。

Juis is designed to meet the daily detox needs of individuals who pay attention to a healthy life but have difficulty in drinking water, and aims to gain the habit of drinking water by sweetening the water. It has a combination of diverse recycled and toxic-free materials.

VESTEL SOUS-VIDE

作者：Murat Hondu/şükran Kasap/Vestel White Industrial Design Team
机构：Vestel Beyaz Eşya San.VeTic.A.Ş
国家：土耳其
组别：产业组

AUTHOR : Murat Hondu/şükran Kasap/Vestel White Industrial Design Team
UNIT : Vestel Beyaz Eşya San.VeTic.A.Ş
COUNTRY : Turkey
GROUP : Product Group

凭借其灵活的制造设施和简化的设计，它扩展了烹饪选项，为用户开辟了新的可能性。它具备卓越的烹饪性能，内嵌了技术改进。设计目标包括改进的用户界面、环保特性、与厨房布局的兼容性以及性能改进的细节。

With its flexible manufacturing facilities and simplistic design, it expands cooking options and opens up new possibilities to the user. It is featuring an excellent cooking performance with technological improvements embedded into it. An improved UI; environmentally-friendly features; compatibility of kitchen layout; performance improving details were the design goals.

苏泊尔无涂层不粘炒锅
SUPOR Ti-G WOK

作者：顾佳俊 臧璐娟 姚新根
机构：武汉苏泊尔炊具有限公司
国家：中国
组别：产业组

AUTHOR : Gu Jiajun, Zang Lujuan, Yao Xin'gen
UNIT : Wuhan SUPOR Cookware Co., Ltd.
COUNTRY : China
GROUP : Product Group

Supor Ti-G 不锈钢炒锅通过钛熔覆技术为消费者提供无高分子物质涂层的健康不粘炒锅。

Supor Ti-G stainless steel work provides users with healthy nonstick work without polymer coating through Titanium alloy coating by Laser Cladding.

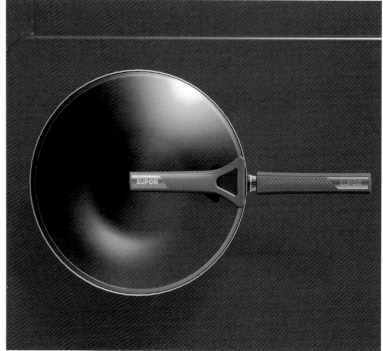

AQUA TEAPOT

BONIQ Pro 2 Sous Vide Cooker

作者 : KERİM KORKMAZ/PELİN DOĞANAY/AYŞE SEÇİL DERELİ İLGİNLİ/
MÜCAHİT BARUĞ/GÖKHAN ŞİMŞEK
机构 : KORKMAZ MUTFAK EŞYALARI SAN. ve TC. A.Ş.
国家 : 土耳其
组别 : 产业组

AUTHOR : KERİM KORKMAZ/PELİN DOĞANAY/AYŞE SEÇİL DERELİ İLGİNLİ/
MÜCAHİT BARUĞ/GÖKHAN ŞİMŞEK
UNIT : KORKMAZ MUTFAK EŞYALARI SAN. ve TC. A.Ş.
COUNTRY : Turkey
GROUP : Product Group

Aqua 是一款多功能茶壶套装，用户可以使用其集成的法式压滤器来泡茶和过滤咖啡。

Aqua is a versatile teapot set that offers the user the ability to brew tea and filter coffee with its integrated French press strainer.

作者 : Kazuhiro HADA/Akihiro MOMOZAKI
机构 : Hayama-Colony Inc.
国家 : 日本
组别 : 产业组

AUTHOR : Kazuhiro HADA/Akihiro MOMOZAKI
UNIT : Hayama-Colony Inc.
COUNTRY : Japan
GROUP : Product Group

BONIQ Pro 2是一款结构紧凑、设计简约但性能高的低温慢煮机。它可以很容易地、美味地、同时大量地烹调蛋白质，使您可以切换到以蛋白质为主的饮食，而不是以碳水化合物为主的饮食。

The BONIQ Pro 2 is a compact, minimalist design but high performance sous vide cooker. Proteins can be cooked easily, tastily and in large quantities simultaneously, allows you to switch to a protein-based diet rather than a carbohydrate-based one.

KITCHEN SCISSORS BLACK DYEING

作者：Takashi Hasegawa/Hossain Arif/Ryota Yamanaka/Shimizu Takashi/Manabu Soeda
机构：Sakaken Co.,Ltd.
国家：日本
组别：产业组

AUTHOR : Takashi Hasegawa/Hossain Arif/Ryota Yamanaka/Shimizu Takashi/Manabu Soeda
UNIT : Sakaken Co.,Ltd.
COUNTRY : Japan
GROUP : Product Group

高品质的"黑染不锈钢厨房剪刀"。这是一款全黑色时尚多功能黑染厨房剪刀。由于这款剪刀采用全不锈钢材质，所以不易生锈，在厨房或户外都能时尚地处理食材。锋利和易用性是这款剪刀的主要特点。

High quality "black-dyed Stainless Steel Kitchen Scissors". It is an all-black stylish multi-functional black-dyed kitchen scissors. Since it is all stainless steel, it does not rust easily and can be handled fashionably in the kitchen or outdoors. The sharpness and the usability are the main features of this Scissors.

小 C 多功能主厨机
"SMALL C" MULT — IFUNCTION COOKING MACHINE

作者：霍磊 成凯 宋业超 杨森
机构：苏泊尔
国家：中国
组别：产业组

AUTHOR : Huo Lei, Cheng Kai, Song Yechao, Yang Sen
UNIT : SUPOR
COUNTRY : China
GROUP : Product Group

小 C 多功能主厨机是一款通过高度整合自动烹饪程序的方式解决用户烹饪困难问题的多功能产品。

"Small C" multi-function cooking machine is a multi-functional product that solves users' cooking difficulties through highly integrated automatic cooking programs.

WOQ

作者：Mauricio Carvajal, Bram Broeken
机构：Recook, Design2Gather
国家：荷兰
组别：产业组

AUTHOR : Mauricio Carvajal, Bram Broeken
UNIT : Recook, Design2Gather
COUNTRY : Dutch
GROUP : Product Group

传统的炒锅烹饪方式，是通过倾斜炒锅将油和香料加热到非常高的温度，这在电磁炉上是做不到的。WOQ 通过特别设计的第二加热面和低压烹饪盖，将炒锅的全部功能带回到现代厨房。

The traditional way of cooking with a wok, heating-up oil and spices to very high temperatures by tilting the pan, is not possible on induction cooktops. WOQ brings back the full versatility of a wok to modern kitchens with the specially designed second heating surface and low-pressure cooking lid.

留同灯
RD LAMP

作者：高晟晏
机构：电子科技大学中山学院
国家：中国
组别：概念组

AUTHOR : Gao Shengyan
UNIT : Zhongshan Institute,University of Electronic Science and Technology
COUNTRY : China
GROUP : Concept Group

升降保温灯对情感化和餐桌空间的高效使用进行探索，通过升降不同的高度对应满足餐桌照明、饭菜保温、食物保鲜的功能。升降保温灯悬挂时打开灯的开关即可满足餐桌照明的需求；轻拉玻璃罩的软胶抓手将自动降下，开始加热并转换为暖光实现饭菜保温；将主体轻按至主体与玻璃罩之间直至出现所需的透气空隙，即可实现食物保鲜。

The height-adjustable thermal lamp explores the efficient utilization of emotions and dining table space. It serves the functions of table lighting, meal warming, and food preservation by adjusting to different heights. When suspended, the lamp can be turned on to provide table lighting. Lightly pulling the soft rubber handle of the glass cover will automatically lower it, initiating the heating function and transforming it into warm light for meal warming. Pressing the lamp's body slightly to create the desired ventilation gap between the body and the glass cover allows for food preservation.

RIPPLING — TOWARD A WIRELESS FRIENDLY ARCHITECTURE

一屏清风 —— 主动交互式动态空间氛围营造曲屏
ONE SCREEN BREEZE — ATMOSPHERE CREATION SCREEN

作者：Kuangxi Cui
机构：Imperial College London
国家：英国
组别：概念组

AUTHOR : Kuangxi Cui
UNIT : Imperial College London
COUNTRY : UK
GROUP : Concept Group

作者：侯嘉璐 林芝 邵义晗 黄艳平
机构：山东工艺美术学院
国家：中国
组别：概念组

AUTHOR : Hou Jialu, Lin Zhi, Shao Yihan, Huang Yanping
UNIT : Shandong University of Art&Design
COUNTRY : China
GROUP : Concept Group

Rippling 是一种建筑／家具材料，可以帮助我们开发适用于 Wi-Fi 友好型架构的电磁家具。该项目试图在一个新的干预点提高无线信号效率，不是信号信道的端点，而是信道本身，即信号传播的环境。

Rippling is a construction/furnish material that could help us develop electromagnetic furniture for a Wi-Fi friendly architecture. The project tries to improve wireless signal efficiency at a novel intervention point, not the end points of a signal channel but the channel itself, the environment in which signals propagate.

"一屏清风"是可以自主检测调节室内呼吸质量和生活氛围的动态屏风，旨在探索未来世界人机交互新关系。

The "One Screen Breeze" is a dynamic screen that can independently detect and adjust the breathing quality and living atmosphere in the room. It aims to explore the new relationship between human and machine interaction in the future world.

追觅精油护发吹风机
DREAME ESSENTIAL OIL HAIR DRYER

作者：姜楠 唐宏旭
机构：追觅创新科技（苏州）有限公司
国家：中国
组别：产业组

AUTHOR : Jiang Nan, Tang Hongxu
UNIT : Dreame Innovation Technology (Suzhou) Co., Ltd.
COUNTRY : China
GROUP : Product Group

追觅精油护发吹风机使用13万转高速无刷电机，并附带精油发膜风嘴，在快速干发的同时柔顺秀发、呵护头皮。

Dreame Hair Care Essence high-speed hair dryer uses a 130,000-rpm high-speed brushless motor and comes with an essential oil hair mask air nozzle, which can soften the hair and care for the scalp while drying the hair quickly.

Air 3 电暖器
AIR 3 HEATER

作者：邱思敏 邱思涛
机构：艾美特电器（深圳）有限公司 / 上海邱思设计咨询有限公司
国家：中国
组别：产业组

AUTHOR : Qiu Simin, Qiu Sitao
UNIT : Airmate / QIU design
COUNTRY : China
GROUP : Product Group

Air3电暖器是一款以"提篮"为设计原型，可以满足取暖、烘衣、杀菌、挂毛巾以及便于移动等需求的电暖器。

We take the 'carrying basket' as the prototype, designed our electric heater that can both heating, drying, sterilizing, tower hanging and is easy to carry.

INCENSMOKE

作者：Eun bin Kim
机构：Hongik University
国家：韩国
组别：概念组

AUTHOR : Eun bin Kim
UNIT : Hongik University
COUNTRY : South Korea
GROUP : Concept Group

"Incensmoke"是一款结合线香与香座功能于一身的加湿器，以解决线香燃烧带来的问题。它比传统的线香更安全，更可持续。

"Incensmoke"is a product that combines the function of incense stick and holder in one humidifier (instead of combustion) to solve the problems caused by the combustion of the incense sticks. It is safer and more sustainable than conventional incense sticks.

Q5 SERIES (Q5/Q5+)

作者：孟繁威 孙凯 于泽晨
机构：北京石头世纪科技股份有限公司
国家：中国
组别：产业组

AUTHOR : Meng Fanwei, Sun Kai, Yu Zechen
UNIT : Beijing Roborock Technology Co., Ltd.
COUNTRY : China
GROUP : Product Group

Roborock 智能扫拖一体机 Q5 series (Q5/Q5+)，吸力升级，配合自动集尘充电桩可实现全自动化的清洁体验。

With upgraded suction power and the automated dust collection docking station, Roborock smart vacuum and mop all-in-one machine Q5 series (Q5/Q5+) achieves a fully automated cleaning experience.

多功能台面吸尘器
MULTI — FUNCTIONAL DESKTOP VACUUM CLEANER

作者：邵荔 杨杰 葛行军 刘前山 刘伟
机构：格力博（江苏）股份有限公司
国家：中国
组别：产业组

AUTHOR : Shao Li, Yang Jie, Ge Xingjun, Liu Qianshan, Liu Wei
UNIT : Globe (Jiangsu) Co., Ltd.
COUNTRY : China
GROUP : Product Group

这是一款秉承以人为本设计理念的吸尘器，创新地将手提包外形融入其中，弱化了工具属性，提升了家居美学。

This is a wireless tabletop vacuum cleaner adhering to the people-oriented design concept, innovatively integrating the shape of the handbag into it, weakening the tool attribute and improving the home aesthetics.

添可智能地毯清洗机
CARPET ONE

作者：何吾佳 王易杰 张昊强
机构：添可智能科技有限公司
国家：中国
组别：产业组

AUTHOR : He Wujia, Wang Yijie, Zhang Haoqiang
UNIT : Tineco Intelligent Technology Co., Ltd.
COUNTRY : China
GROUP : Product Group

Carpet One 是一台具有烘干功能，适用不同场景的智能地毯清洗机。它开创式地增加了烘干功能，用户能方便地用热风烘干地毯，地毯的干燥程度会显示在 LCD 屏幕上，方便用户调整烘干时间。同时，本产品创新地配备了 iLoop 传感器，可通过屏幕显示当前地毯的脏污程度，满足用户对了解地毯状态的需求。

Carpet One is an intelligent carpet cleaning machine with drying functionality, suitable for various scenarios. It introduces an innovative drying feature, allowing users to conveniently dry carpets using hot air. The carpet's dryness level is displayed on the LCD screen, enabling users to adjust the drying time easily. Additionally, this product comes equipped with the iLoop sensor, which can display the current level of carpet dirtiness on the screen, meeting users' needs for understanding the carpet's condition.

它医生猫砂盆
CAT LITTER BASIN

作者：李细强 方兆杰 王建仁 王国强
机构：上海六分之一宠物用品有限公司 / 宁波法诺工业产品设计有限公司
国家：中国
组别：产业组

AUTHOR : Li Xiqiang, Fang Zhaojie, Wang Jianren, Wang Guoqiang
UNIT : Shanghai One Sixth Pet Products Co., Ltd. / Ningbo Fonu Industrial Product Design Co., Ltd.
COUNTRY : China
GROUP : Product Group

这款猫砂盆通过搭载创新的"新风模块"有效解决猫砂盆易产生异味的问题，为养宠人士提供舒心健康的环境。

Equipped with a creative"fresh air module", the litter box can effectively reduce the odor, providing a comfortable and healthy environment for cat parents.

悬浮空气扇
SUSPENDED FAN

作者：卢传德 杨扬 杨能鹏 罗玮瑜
机构：广东顺德米壳工业设计有限公司
国家：中国
组别：产业组

AUTHOR : Lu Chuande, Yang Yang, Yang Nengpeng, Luo Weiyu
UNIT : MIKO Industrial Design Co.,Ltd.
COUNTRY : China
GROUP : Product Group

这是一款能净化的负离子3D循环风扇。产品采用纳米离子净化除菌，让空气时刻保持洁净，足不出户感受自然风。

This is a purify anion 3D circulating fan. Nano ion purification and sterilization, so that the air always keeps clean, stay indoors to feel the natural wind.

DOUBLE SIDED CHAIR

CIRCO

作者 : Fang Li, Yue Chen, Jun Wang, Mingdong, Yining Xu
机构 : Inner Mongolia Normal University / Junwang Studio
国家 : 意大利
组别 : 概念组

AUTHOR : Fang Li, Yue Chen, Jun Wang, Mingdong, Yining Xu
UNIT : Inner Mongolia Normal University / Junwang Studio
COUNTRY : Italy
GROUP : Concept Group

作者 : Burak Emre ALTINORDU, Kiraz Sema TURHAN, Can UCKAN YUKSEL
机构 : Vestel Electronics
国家 : 土耳其
组别 : 产业组

AUTHOR : Burak Emre ALTINORDU, Kiraz Sema TURHAN, Can UCKAN YUKSEL
UNIT : Vestel Electronics
COUNTRY : Turkey
GROUP : Product Group

双面椅是一种多用途的家具。它既可以用作客厅椅, 也可以用作边桌。由于结构设计巧妙, 只需翻转椅子靠背即可改变家具的功能。它具有独特的美感, 材质为铁管, 耐腐蚀、耐磨, 适合室内外使用。将椅子靠背翻转180°, 可完成椅子与边桌的功能转换。这个功能非常适合习惯于高度灵活性的现代家庭。

Dual-Purpose Chair is a versatile piece of furniture that can be used both as a living room chair and as a side table. Thanks to its clever structural design, you can easily change the function of the furniture by flipping the chair's backrest. It boasts a unique aesthetic, made from corrosion-resistant and durable iron tubing, suitable for both indoor and outdoor use. Simply flip the chair's backrest 180 ° to transform it from a chair into a side table. This feature is particularly well-suited for modern households that value flexibility.

Circo 是一款外形优雅独特的电视遥控器。在满足用户的人体工程学和动态形式的同时, 它强调了其图标形状的按钮和银黑色调的现代设计理念。

Circo is a TV remote control with an elegant and unique form. While satisfying the user with its ergonomic and dynamic form, it emphasizes the modern design concept with its icon shaped buttons and silver & black tones.

森友汇 W1 擦窗机器人适配吸尘器
W1 WINDOW CLEANING ROBOT (CONFIGURABLE VACUUM)

作者：赵婧 谭欣怡 赵晋霞 郭少轩 潘思博
机构：山西阿迈森机器人技术有限公司
国家：中国
组别：产业组

AUTHOR : Zhao Jing, Tan Xinyi, Zhao Jinxia, Guo Shaoxuan, Pan Sibo
UNIT : Shanxi Ameson Robot Technology Co., Ltd.
COUNTRY : China
GROUP : Product Group

世界首款搭配吸尘器使用的擦窗机，依靠吸尘器吸力，实现碎片化时间的玻璃清洁自由，满足快速擦窗需求。

The world's first window cleaner with a configured vacuum cleaner, relying on the suction power of vacuum cleaner to achieve window cleaning in fragmentation time faster and safer.

童车可持续产品 - 服务系统
CIRCU

作者：尹书乐 张泽航 张晓慧 罗博源
机构：杭州深氪创意设计有限公司
国家：中国
组别：概念组

AUTHOR : Yin Shule, Zhang Zehang, Zhang Xiaohui, Luo Boyuan
UNIT : Hangzhou Shenkron Creative Design Co.,Ltd.
COUNTRY : China
GROUP : Concept Group

circu 是一款通过婴儿车、学步车、滑板车的结构转换的方式，解决了传统童车产品功能单一、替换率高、使用周期短的问题。

circu is a way to convert the structure of the baby carriage walker scooter, solving the problem of high replacement rate of traditional stroller products with a single function and short use cycle.

儿童遛娃神器
LIGHT WEIGHT STROLLER

作者：薛瑞德
机构：宁波鸭嘴兽贸易有限公司
国家：中国
组别：产业组

AUTHOR : Xue Reide
UNIT : NINGBO PLATYPUS TRADING CO.,LTD.
COUNTRY : China
GROUP : Product Group

这是一款专门为宝妈解决带娃出行难题，质量可靠的儿童推车。产品轻便小巧可折叠，工艺精湛，结构稳固，安全性高。市面上的婴儿推车可靠性高的普遍比较笨重，轻便的品质又比较低劣，我们的产品正好填补了这个空白，让宝妈解放双手，轻松、安全、可靠地带娃出行。

This is a children's stroller specifically designed for busy moms, providing a compact and foldable solution with high-quality and reliable craftsmanship. It's lightweight, compact, and foldable, featuring excellent craftsmanship and a sturdy structure, ensuring high levels of safety and reliability. Many baby strollers on the market are either reliable but bulky or lightweight but of lower quality. Our product fills this gap, allowing moms to free their hands and travel with their children easily, safely, and reliably.

象拓智能爬梯轮椅
ELEFANTO

作者：姜天明 郑世仁 谭陽熙 邱鹏洋 林炫燚
机构：黎沛特科技
国家：中国
组别：产业组

AUTHOR : Jiang Tianming, Zheng Shiren, Tan Yangxi, Qiu Pengyang, Lin Xuanyi
UNIT : Libpet Tech Limited
COUNTRY : China
GROUP : Product Group

Elefanto 是一款为行动不便的用户提供的人工智能爬楼梯轮椅，由人工智能传感器融合模型和变形机械设计所赋能。

Elefanto is an AI stair climbing wheelchairs for users with challenged mobility, empowered by our AI sensor fusion model and deformable mechanical design.

STROLLEAZI — REVOLUTIONISING THE MODERN-DAY TRAVEL SYSTEM FOR PARENTS WORLDWIDE

INFORM

作者：Emma Lawrence, Jesper Tyrer, Adrian Batchelor, Shaun Carine, Stacy Marshall
机构：Strolleazi UK Ltd.
国家：英国
组别：产业组

AUTHOR : Emma Lawrence, Jesper Tyrer, Adrian Batchelor, Shaun Carine, Stacy Marshall
UNIT : Strolleazi UK Ltd.
COUNTRY : UK
GROUP : Product Group

作者：Rick Van Schie
机构：Eindhoven University of Technology
国家：荷兰
组别：概念组

AUTHOR : Rick Van Schie
UNIT : Eindhoven University of Technology
COUNTRY : Dutch
GROUP : Concept Group

Strolleazi 的一体化设计，让父母不再需要购买仅解决儿童特定阶段问题的多个独立模块的婴儿车和汽车座椅。Strolleazi 将多个不同的产品整合到一个简单、单一的体验中，陪同您的孩子一起成长：一辆婴儿车和一个汽车安全座椅。

Strolleazi's integrated design removes the need for parents to purchase multiple separate modules of prams & car seats, which only solve for a specific stage of childhood. Strolleazi consolidates multiple disparate products into a simple, single experience that grows with your child; one pram & one car seat.

InForm 是一套摩托车握把，可提高骑手的情境感知能力。握把会改变形状以指出盲点或碰撞警告，从而帮助骑手更快更准确地对危险作出反应。

InForm is a set of motorcycle grips that improve the rider's situational awareness. The grips change shape to point out blind spot or collision warnings, thereby helping the rider react quicker and more accurately to hazards.

唯蜜瘦便携式智能腹部按摩仪
VMESHOU PORTABLE ABDOMINAL INTELLIGENT MASSAGER

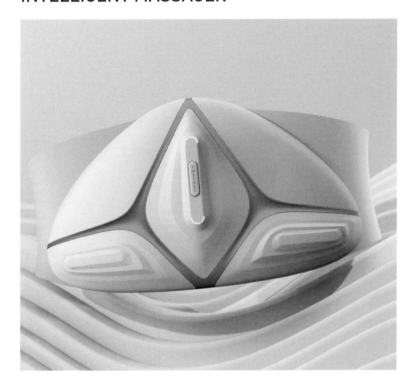

作者: 陆明豪 吴梦璇 潘浩男 张铁澄 胡羽
机构: 湖南旻一科技有限公司 / 杭州热浪创新控股有限公司
国家: 中国
组别: 产业组

AUTHOR : Lu Minghao, Wu Mengxuan, Pan Haonan, Zhang Tiecheng, Hu Yu
UNIT : Hunan Minyi Technology Co.,Ltd. / Hangzhou Hotdesign group Co.,Ltd.
COUNTRY : China
GROUP : Product Group

唯蜜瘦便携式智能腹部按摩仪,开创性融合中医药温灸法与 EMS 脉冲智能科技,以舒适体验实现科学的身材管理。

VMESHOU Portable Abdominal Intelligent Massager,the pioneering integration of traditional Chinese medicine warm moxibustion and EMS pulse intelligent technology to achieve scientific body management with comfortable experience.

ALEXIA AUDRAIN

作者: Audrain Alexia, Lemaitre Corentin
机构: LABAA
国家: 法国
组别: 产业组

AUTHOR : Audrain Alexia, Lemaitre Corentin
UNIT : LABAA
COUNTRY : France
GROUP : Product Group

专为自闭症患者设计,每个人都可以使用。这款创新的椅子具有可充气的内壁,可以对身体施加深层压力。这种压力有助于减轻感觉障碍患者的焦虑,因此让他们感觉更平静,更有时间,能够参与学习或社交互动。

Designed for people with autism, usable by everyone.This innovative chair has inflatable inner walls that apply deep pressure on the body. This pressure helps reduce anxiety in people with sensory disorders, so they can feel calmer, more available and able to participate in learning or social interaction.

BOOST ——儿童辅助装置
BOOST — CHILD ASSIST DEVICE

作者：郑斌
机构：深圳市意臣工业设计有限公司
国家：中国
组别：产业组

AUTHOR : Zheng Bin
UNIT : Innozen Product Design Co., Ltd.
COUNTRY : China
GROUP : Product Group

该辅助装置专为成长中的青少年（6~18岁）设计，通过符合人体工程学的关节设置满足用户日常抓握需求和装饰需求。

Designed for growing teenagers (6-18 years old), the assistive device meets the user's daily grasping needs and decorative needs through ergonomic joint settings.

雅威 PRO 300
AAVI PRO 300

作者：黄将 侯壮 刘思亮
机构：雅威科技有限公司 / 北京木马工业设计有限公司
国家：中国
组别：产业组

AUTHOR : Huang Jiang, Hou Zhuang, Liu Siliang
UNIT : Aavi Technologies Ltd. / Beijing Moma Industrial Product Design Co.,Ltd.
COUNTRY : China
GROUP : Product Group

AAVI Pro 300力求为每位用户提供一个安全健康的呼吸环境而设计。

AAVI Pro 300 strives to provide each user with a safe and healthy breathing environment.

ASP 共享机车安全帽
ASP - HELMET FOR SHARED SCOOTERS

扣扣就走出国用婴儿载具
COCOGO

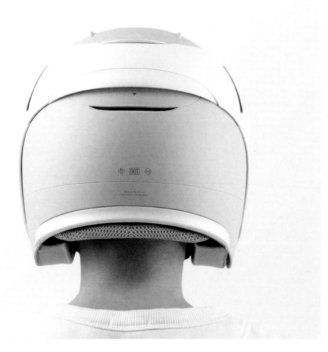

作者：陈俞安
国家：中国（台湾）
组别：概念组

AUTHOR : Chen Yu'an
COUNTRY : Taiwan, China
GROUP : Concept Group

作者：范競文
国家：中国（台湾）
组别：概念组

AUTHOR : Fan Jingwen
COUNTRY : Taiwan, China
GROUP : Concept Group

ASP 通过改变现有安全帽的内衬材料，来解决共享机车安全帽的卫生与清洁问题，同时创造更舒适的移动体验。

ASP is a concept that solved the problem of hygiene and cleanliness of shared motorcycle helmets by changing the lining material of shared scooter helmets and creating a more comfortable mobility experience.

COCOGO 是一款通过结构、造型调整使父母与婴儿出国游玩于各种环境时，拥有更便捷移动方案的婴儿载具产品。

COCOGO is a baby carrier product that enables parents and babies to have a more convenient mobility solution when playing abroad in various environments by adjusting the structure and shape.

SMART BASSINET

手持消毒喷雾
HAND-HELD DISINFECTANT SPRAY

作者： Shin Jung A, Park Geon Hee, Kim Ki Yong, Lee Yu Jin, Seo Hae Sung
机构： Design Uno
国家： 韩国
组别： 产业组

AUTHOR : Shin Jung A, Park Geon Hee, Kim Ki Yong, Lee Yu Jin, Seo Hae Sung
UNIT : Design Uno
COUNTRY : South Korea
GROUP : Product Group

作者： 王权一 卢传德 杨能鹏 杨扬 罗玮瑜
机构： 广东顺德米壳工业设计有限公司
国家： 中国
组别： 产业组

AUTHOR : Wang Quanyi, Lu Chuande, Yang Nengpeng, Yang Yang, Luo Weiyu
UNIT : MIKO Industrial Design Co.,Ltd.
COUNTRY : China
GROUP : Product Group

智能摇篮为婴儿提供最佳的睡眠环境，减轻育儿压力，改善育儿环境。它配备了白噪声和全方位通风结构。为了防备意外情况，它提供了可移动检查的监测系统和风险检测警报，以确保在父母忙于家务时婴儿的安全。通过各种传感器检测身高、体重和排便情况，并提供给家长。

The Smart Cradle provides the optimal sleep environment for babies, reducing parenting stress and improving the overall parenting environment. It is equipped with white noise and a 360-degree ventilation system. To prevent unforeseen situations, it offers a movable inspection monitoring system and risk detection alerts to ensure the baby's safety while parents are busy with household chores. It detects height, weight, and bowel movement using various sensors and provides this information to parents.

这是一款低成本、易携带的手持消毒喷雾，将盐和水倒入容器，通过电解产生氨酸消毒液，便能快速进行消毒。

This is a low-cost, portable, hand-held disinfecting spray that is quickly disinfected by pouring salt and water into a container and producing an acid disinfectant by electrolysis.

梵品小木马椭圆机
FAMIFIT ELLIPTICAL MACHINE

作者 : 高宇玄
机构 : 布梵（上海）健康科技有限公司
国家 : 中国
组别 : 产业组

AUTHOR : Gao Yuxuan
UNIT : Bufan (Shanghai) Health Technology Co., Ltd.
COUNTRY : China
GROUP : Product Group

梵品小木马椭圆机，弹力棒椭圆机品类开创者，立志于开发更符合家庭使用的运动产品。

FamiFit elliptical machine, pioneer of elliptical machine with flexi-bars. Created a new generation of sports equipment that is more suitable for home use.

WELME

作者 : Rahul Chopra, Umang Shah
机构 : Camex Wellnest Ltd. Oro Innovations Pvt. Ltd.
国家 : 印度
组别 : 产业组

AUTHOR : Rahul Chopra, Umang Shah
UNIT : Camex Wellnest Ltd. Oro Innovations Pvt. Ltd.
COUNTRY : India
GROUP : Product Group

Welme 经期痉挛缓解器是专为缓解女性经期疼痛而设计的，适用于缓解无法停止的经期痉挛。这是一款功能强大的设备，低调、美观、适合女性使用，采用 TENS 技术，已知可通过刺激神经来阻断疼痛信号，且无副作用，确保女性的安全系数。

Welme period cramp machine is designed to relieve period pain for women with unstoppable cramps. A powerful device that is discreet, aesthetically sound, women-friendly and comprises of TENS technology known to block the pain signals by stimulating nerves with no side effects, ensuring the safety factor for women.

COURAGE — ECG HEALTH KIT FOR KIDS

NETSU!

作者：Jun Wang, Xia Hua, Lingxiao Zeng, Mingdong Mao, Yining Xu
机构：Junwang Studio
国家：意大利
组别：产业组

AUTHOR : Jun Wang, Xia Hua, Lingxiao Zeng, Mingdong Mao, Yining Xu
UNIT : Junwang Studio
COUNTRY : Italy
GROUP : Product Group

作者：夏骄阳
国家：中国
组别：概念组

AUTHOR : Xia Jiaoyang
COUNTRY : China
GROUP : Concept Group

这是一款专为儿童设计的无线心电图（EGG）测量设备。无线心电图仪的导联头结构紧凑，使用方便，避免了电线带来的麻烦。整个测量过程设计成游戏形式，心电图徽章象征着获奖荣耀，连接心电图电极的行为被伪装成颁发奖牌。完成后取下心电图徽章，交给孩子作为荣誉勋章保存。

This is a wireless electrocardiogram (ECG) measuring device designed specifically for children. The compact electrode structure of the wireless ECG device is user-friendly and eliminates the hassle of wires. The entire measurement process is designed in the form of a game, where the ECG badge symbolizes the honor of winning, and the act of connecting the ECG electrodes is disguised as presenting a medal. After completion, the ECG badge is removed and given to the child to be kept as an honorary medal.

"NETSU!" 是一款基于温敏变色材料设计的体温计。在手持端，设计有"37.5℃"与内含"！"字符的红色圆形图标，意在通过发热温度与警示颜色的关联性表示，对用户的测试结果进行提示。温度计的造型思路沿用传统水银温度计的整体尺寸、形态、使用方式，遵循大众对测温行为的习惯性记忆，降低用户对于体温计的陌生感与学习成本，使用户能够将关注点完全围绕于对新型材料的感知。

"NETSU!"is a thermometer designed based on temperature sensitive color changing materials.On the handheld end, you'll find a red circular icon with a "37.5 ℃ " mark and an embedded "!" character. This design is intended to provide users with clear temperature test results by linking the heating temperature with warning colors. The thermometer's design concept follows the traditional size, shape, and usage of mercury thermometers, in line with the familiar habits of temperature measurement for the general public. This approach minimizes any sense of unfamiliarity and learning curve for users, allowing them to focus solely on experiencing the innovative material.

WE | AVER+

BLOOD PRESSURE METER FOR KIDS

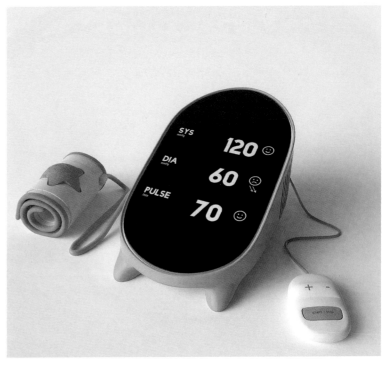

作者 : Yumeng Li, Zongheng Sun
机构 : LI&SUNDESIGN LLC; PEAR & MULBERRY
国家 : 美国
组别 : 概念组

AUTHOR : Yumeng Li, Zongheng Sun
UNIT : LI&SUNDESIGN LLC; PEAR & MULBERRY
COUNTRY : United States
GROUP : Concept Group

Welaver+ 是一款由可回收 TPU 制成的治疗鞋，为足部生长迅速的8~14岁儿童提供足部护理，以保护他们免受慢性足跟不适和跟骨骨突炎的困扰并使其康复。3D 打印模块参数化地实现了支撑 - 舒适的平衡结构，而不会妨碍足部的正常生长。

Welaver+, as therapeutic shoes made from recyclable TPU, provide foot care for children aged 8-14 whose feet grow rapidly, to protect and rehabilitate them from chronic heel discomfort and calcaneal apophysitis. The 3D-printed modules parametrically achieve a support-comfort balanced structure without hindering the normal growth of the foot.

作者 : Jun Wang, Yue Chen, Mingxi Sun, Mingdong Mao, Yining Xu
机构 : Junwang Studio
国家 : 意大利
组别 : 概念组

AUTHOR : Jun Wang, Yue Chen, Mingxi Sun, Mingdong Mao, Yining Xu
UNIT : Junwang Studio
COUNTRY : Italy
GROUP : Concept Group

这是一套用于检测儿童血压的医疗设备，可减少儿童的恐惧心理。使用时袖带充气，压力逐渐增加。最后，一个漂亮的黄色小五角星就会出现。它能极大地吸引孩子的注意力，减轻孩子的压力。充气部分由可回收织物制成，可以定制不同的图案。

This is a medical device designed for monitoring children's blood pressure, aiming to reduce children's anxiety during the process. When in use, the cuff inflates gradually, exerting pressure. Eventually, a charming yellow pentagram appears, captivating the child's attention and alleviating their stress. The inflatable component is crafted from recyclable fabric, allowing for customizable designs to further engage children.

可穿戴式 —— 便携式输液袋
WEARABLE — PORTABLE INFUSION BAG

偏移创可贴 —— 用14mm 解决问题
OFFSET BAND-AID — SOLVING PROBLEMS WITH 14MM

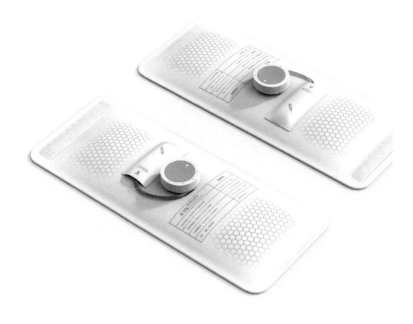

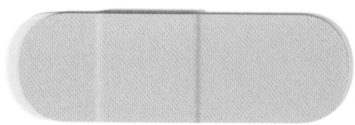

作者：韦海东 李博洋 赵剑锋
国家：中国
组别：产业组

AUTHOR：Wei Haidong, Li Boyang, Zhao Jianfeng
COUNTRY：China
GROUP：Product Group

作者：张迎丽
机构：湖南大学
国家：中国
组别：概念组

AUTHOR：Zhang Yingli
UNIT：Hunan University
COUNTRY：China
GROUP：Concept Group

便携式输液袋是一款适用于室内外特殊场景（急救现场，救灾）使用的医疗产品。

Portable infusion bag is a medical product suitable for special indoor and outdoor scenarios (emergency scene, disaster relief).

此偏移创可贴是根据人们从一侧使用胶带粘贴的习惯而设计的一款创可贴。

This offset band-aid is a band-aid designed according to people's habit of using adhesive tape from one side.

CIRCULAR LIGHT

重复利用快递包装设计
REUSE EXPRESS PACKAGING DESIGN

作者 : Mariet Schreurs, Toros Cangar, Xiuwen Liao, Zhiguang He, Hui Chen
机构 : Orange Creatives
国家 : 荷兰
组别 : 概念组

AUTHOR : Mariet Schreurs, Toros Cangar, Xiuwen Liao, Zhiguang He, Hui Chen
UNIT : Orange Creatives
COUNTRY : Dutch
GROUP : Concept Group

作者 : 张振宇
国家 : 中国
组别 : 概念组

AUTHOR : Zhang Zhenyu
COUNTRY : China
GROUP : Concept Group

循环灯是在考虑循环经济和碳足迹的情况下打造的。采用模块化设计，易于组装、拆卸和维修，可平装运输，使产品维护变得高效、便捷。该产品主要由各种回收材料制成，这些材料来自现有的塑料瓶、瓶盖、金属罐和木材。使用回收塑料、金属和木材赋予了它一种再生和共生的感觉。

The Circular Light was created with consideration of the Circular Economy and Carbon Footprint. It is modular in design, easy to assemble, disassemble and repair; flat-packed for transport. Due to the modular structure and simple DIY assembly method, the product maintenance becomes efficient and convenient. The product is predominately made from various recycled materials accumulated from existing plastic bottles, caps, metal cans and wood. The use of recycled plastics, metals and wood, gives it a sense of regeneration and symbiosis.

我国每年产生大量的快递包装垃圾，为了解决这个问题设计了一款可以重复利用的快递包装。

My country produces a lot of express packaging waste every year. In order to solve this problem, a reusable express packaging is designed.

户外电源
OUTDOOR POWER

作者：刘杨 刘川胜 赵同心
机构：首辅锂电池科技江苏有限公司 / 苏州上品极致产品设计有限公司
国家：中国
组别：产业组

AUTHOR : Liu Yang, Liu Chuansheng, Zhao Tongxin
UNIT : Shoufu Lithium Battery Technology Jiangsu Co., Ltd. / Suzhou Top Industrial Design Co., Ltd.
COUNTRY : China
GROUP : Product Group

可靠的 BMS 保护便携式电站的安全运行并快速响应，适用于户外和室内活动的便携式电源。

Reliable BMS protecting the safe operation of the portable power station and responds quickly, Portable power for both outdoor and indoor activities.

潜水艇旋风式厨余垃圾破碎机
SUBMARINE WHIRLWIND KITCHEN WASTE CRUSHER

作者：唐孝康 陈丹蓉 蔡旭裕 颜艺琳 黄伟军
机构：福建宏讯电子有限公司 / 泉州市数坊信息科技有限公司
国家：中国
组别：产业组

AUTHOR : Tang Xiaokang, Chen Danrong, Cai Xuyu, Yan Yilin, Huang Weijun
UNIT : HIROJIN ELECTRONIC / Quanzhou Sufang Information Technology Co.,Ltd.
COUNTRY : China
GROUP : Product Group

这是一款行业内创新的家用厨余垃圾处理器，产品采用先进的直流无刷马达控制技术，巧妙安排内部组件的布局，模块化设计，从而实现其体积小于80% 的同类产品。独特的潜水艇式设计，简单直接的人机交互，实现功能、技术和外观的统一，时刻保持厨房清洁和卫生，提升用户生活幸福感。

This is an innovative household kitchen waste processor in the industry. The product adopts advanced DC brushless motor control technology, ingeniously arranges the layout of internal components, and adopts modular design, so as to achieve similar products with a volume of less than 80%. The unique submarine design, simple and direct human-computer interaction, achieve the unity of function, technology and appearance, keep the kitchen clean and sanitary at all times, and improve the happiness of users.

COCOHELI

作者: AUTHENTIC JAPAN TEAM
机构: AUTHENTIC JAPAN Co.,Ltd.
国家: 日本
组别: 产业组

AUTHOR : AUTHENTIC JAPAN TEAM
UNIT : AUTHENTIC JAPAN Co.,Ltd.
COUNTRY : Japan
GROUP : Product Group

Cocoheli 是一项"尽早发现山区遇险人员的服务"。当某个人员遇险时，配备专用接收器的直升机会搜索该人员拥有的发射器。

Cocoheli is a "service for early detection of people in distress in the mountains. When a distress occurs, a helicopter equipped with a dedicated receiver searches for the transmitter owned by the member.

助步机
WALKER

作者: 郑斌
机构: 深圳市意臣工业设计有限公司
国家: 中国
组别: 产业组

AUTHOR : Zheng Bin
UNIT : Innozen Product Design Co., Ltd.
COUNTRY : China
GROUP : Product Group

这是一款外骨骼腿部助力机器人，穿戴于人体的腰胯部位，为腿部力量弱以及有步态缺陷的人提供辅助动力。

This is an exoskeleton leg-assisted robot, which is worn on the waist and crotch of the human body to provide auxiliary power for people with weak legs and gait defects.

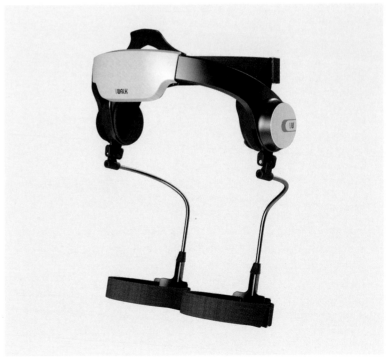

防冲击手套
IMPACT RESISTANT GLOVES

LUDWIG KATRIN 颈部固定器
LUDWIG KATRIN NECK FIXER

作者：赵智峰 唐聪智 陈亚新 芮达志 李得阳
机构：无锡新亚安全用品有限公司 / 苏州柒整合设计有限公司
国家：中国
组别：产业组

AUTHOR : Zhao Zhifeng, Tang Congzhi, Chen Yaxin, Rui Dazhi, Li Deyang
UNIT : Wuxi Safety Products Co.,Ltd. / Suzhou CHY Integration Design Co., Ltd.
COUNTRY : China
GROUP : Product Group

作者：章少伟 邹斯乐 林鸿泰 沈伟杰
机构：广州纽得赛生物科技有限公司
国家：中国
组别：产业组

AUTHOR : Zhang Shaowei, Zou Sile, Lin Hongtai, Shen Weijie
UNIT : GuangZhou New-Design Biotechnology Co.,Ltd.
COUNTRY : China
GROUP : Product Group

这款防护手套为油田钻井平台、建筑工人、地震等灾害救援、执法部门而设计。

Impact resistant gloves are designed for oil field drilling platforms, construction workers, disaster rescue and law enforcement departments.

首款内嵌气管颈部固定器，提供创新的一体充放气使用体验，产品具有档位调整声音反馈。

The first inline airway neck fixator.Innovative integrated inflation and deflation experience.Sound feedback with gear adjustment.

正浩48V 全场景电源系统
ECOFLOW POWER KITS

作者：王雷 郑海斌 郭晨 高伟彬 张一帆
机构：深圳市正浩创新科技股份有限公司
国家：中国
组别：产业组

AUTHOR : Wang Lei, Zheng Haibin, Guo Chen, Gao Weibin, Zhang Yifan
UNIT : EcoFlow Inc.
COUNTRY : China
GROUP : Product Group

正浩48V 全场景电源系统旨在打造简单、安全高效的用电体验，目标成为引领全球房车、小屋等用户群体的离网用电选择，其颠覆了传统离网电源系统设计，以48V 系统 +48V 智能电池全用电生态覆盖，高度集成各个系统模块，融合科技与美学的设计以及智能体验，将简单、安全、高效、有趣的用电体验最大化呈现给用户。

The EcoFlow Power Kits aims to create a simple, safe and efficient power consumption experience, and aims to lead the off-grid power consumption choice for user groups such as RVs, cabins around the world. The EcoFlow Power Kits simplify the demanding installation process of existing custom power solutions, allowing users to tailor their own settings that best fit their power needs with compact and smart modules, whether they are in an RV, yacht, off-grid or retrofitting their workshops. With a unique 48V system, which allows the EcoFlow Power Kits to enhance the user experience on multiple levels including safety, energy efficiency, versatility, and ease of use.

PARK, GROVE AND URBAN FOREST
DESCRIPTIVE UNIT

作者：Ezgi ÖZCAN, Gürkay AYDOĞMUŞ, Koray GELMEZ, Suat Batuhan ESİRGER, Sarp SUSÜZER Ahmet Furkan KELEŞ
机构：iSTON stanbul Concrete Elements / Ready Mixed Concrete Factories Corporation Istanbul Metropolitan Municipality
国家：土耳其
组别：产业组

AUTHOR : Ezgi ÖZCAN, Gürkay AYDOĞMUŞ, Koray GELMEZ, Suat Batuhan ESİRGER, Sarp SUSÜZER Ahmet Furkan KELEŞ
UNIT : iSTON stanbul Concrete Elements / Ready Mixed Concrete Factories Corporation Istanbul Metropolitan Municipality
COUNTRY : Turkey
GROUP : Product Group

本产品带有属于伊斯坦布尔大都市的公园、小树林和城市森林的名称；通过其面板提供信息，并通过其垂直结构和内部照明直观地指示这些区域。

This product bears the name of the parks, groves, and urban forests belonging to the Istanbul Metropolitan Municipality; informs via its panel and visually indicates the areas with its vertical structure and interior lighting.

WATALK TRAVEL

作者 : Shin Jung A, Kim Young Dug, Kim Ki Yong, Seo Hae Sung, Park Geon Hee
机构 : İSTON İstanbul Concrete Elements and Ready Mixed Concrete Factories Corporation
国家 : 韩国
组别 : 产业组

AUTHOR : Shin Jung A, Kim Young Dug, Kim Ki Yong, Seo Hae Sung, Park Geon Hee
UNIT : İSTON İstanbul Concrete Elements and Ready Mixed Concrete Factories Corporation
COUNTRY : South Korea
GROUP : Product Group

这是一款不倒翁式水质测量装置，便于携带，可随时随地使用。

It is a tumbler-type water quality measuring device that can be easily carried and used anytime, anywhere.

产业装备
INDUSTRIAL EQUIPMENT

CELLQUA

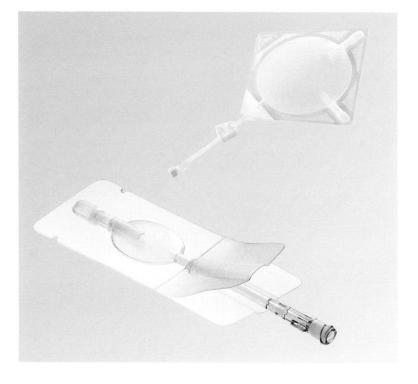

作者：Satoshi Hirai, Makoto Shibuya, Koji Suzuki, Yutaka Katsuno, Shogo Yanagida
机构：JMS Co,.Ltd.
国家：日本
组别：产业组

AUTHOR : Satoshi Hirai, Makoto Shibuya, Koji Suzuki, Yutaka Katsuno, Shogo Yanagida
UNIT : JMS Co,.Ltd.
COUNTRY : Japan
GROUP : Product Group

CellQua 是一种用于冷冻保存收集和培养的人体细胞的容器，主要用于再生医学。CellQua 所采用的材料和形状解决了传统产品的细菌污染、超低温和冷冻状态下造成的破损以及由于残留液体造成的细胞损失等问题。

CellQua is a container used to cryopreserve collected and cultivated human cells, mainly in regenerative medicine. Its material and shape solve the problems of conventional products such as germs contamination, breakage caused under ultra-low temperature and frozen-state, and loss of cells due to the residual liquid.

HELLOFACE TRANSPARENT RESPIRATOR MASK (P3)

作者：Dean Ezekiel, Jamie Ezekiel, Ross Inanc
机构：Handan Hengyong / oneoverthree
国家：英国
组别：产业组

AUTHOR : Dean Ezekiel, Jamie Ezekiel, Ross Inanc
UNIT : Handan Hengyong / oneoverthree
COUNTRY : UK
GROUP : Product Group

Helloface P3口罩通过透明的、兼容人脸识别、经过认证的呼吸器口罩，解决了在有害环境中使用防护口罩进行唇语阅读、面部识别和有效沟通的问题。

Helloface P3 mask solves the problem of lip reading, facial recognition and communicating effectively with protective face masks in harmful environments, by the means of a transparent, face ID compatible, certified respirator mask.

大圣磁控胶囊式内窥镜
DASHENG CAPSULE ENDOSCOPE

BENEFUSION N 智能输液系统
BENEFUSION N SERIES

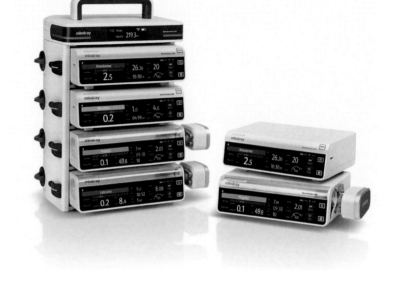

作者：王建平 张晓艳 彭国会
机构：深圳市资福医疗技术有限公司
国家：中国
组别：产业组

AUTHOR : Wang Jianping , Zhang Xiaoyan, Peng Guohui
UNIT : Shenzhen Zifu Medical Technology Co., Ltd.
COUNTRY : China
GROUP : Product Group

作者：何丽娟 柴海波 陈大兵 章蕾 邹小玲
机构：深圳迈瑞科技有限公司 / 深圳迈瑞生物医疗电子股份有限公司
国家：中国
组别：产业组

AUTHOR : He Lijuan, Chai Haibo, Chen Dabing, Zhang Lei, Zou Xiaoling
UNIT : Shenzhen Mindray Technology Co., Ltd. / Shenzhen Mindray Bio-Medical Electronics Co.,Ltd.
COUNTRY : China
GROUP : Product Group

大圣磁控胶囊胃镜真正实现无创无痛无麻醉的胃镜检查，是能够对人体胃部进行精准检查的胶囊胃镜。

Dasheng magnetron capsule gastroscope truly realizes non-invasive, painless and anesthesia-free gastroscopy, and is a capsule gastroscope that can accurately examine the human stomach.

BeneFusion n Series 智能输液系统搭载智慧药物库，以病人安全用药为核心，为临床医护人员打造更安全高效的智能输液系统。

BeneFusion N Series is an intelligent infusion delivery system features the infusion pump with smart drug library. It is designed for clinical staff with safe medication for patients as its core. While ensuring safe medication, this smart infusion tool makes the operation of clinical staff more intuitive and efficient.

COOPERATION ROBOT AND EXCLUSIVE AMR

视觉搬运机器人
VISUAL HANDLING ROBOT

作者：Shin Jung A, Park Jong hoon, Park Geon Hee, Kim Ki Yong, Lee Yu Jin
机构：NEUROMEKA / DESIGNUNO
国家：韩国
组别：产业组

AUTHOR : Shin Jung A, Park Jong hoon, Park Geon Hee, Kim Ki Yong, Lee Yu Jin
UNIT : NEUROMEKA / DESIGNUNO
COUNTRY : South Korea
GROUP : Product Group

作者：何晶杰 叶晓俊 杭航 郑超 徐永奎
机构：杭州蓝芯科技有限公司
国家：中国
组别：产业组

AUTHOR : He Jingjie, Ye Xiaojun, Hang Hang, Zheng Chao, Xu Yongkui
UNIT : Hangzhou Lanxin Technology Co.,Ltd.
COUNTRY : China
GROUP : Product Group

"Moby" 是 Neuromeka 为 "Indy" 开发的自主移动机器人平台。"Moby" 使 "Indy" 拥有不受限制的工作空间。"Moby" 可以通过更换传感器板来配备各种传感器。"Moby" 可以通过更换工作托盘来用于送货、巡逻、检疫和引导。

"Moby" is Neuromeka's autonomous mobile robot platform for "Indy"."Moby" makes "Indy" has a non-restriction workspace. "Moby" can be equipped with various sensors by changing the sensor plate. "Moby" can be used for delivery, patrol, quarantine, and guidance by replacing workpallets.

产品采用了自主开发的 LX-MRDVS 系统，能够准确识别障碍物并绕开障碍物，安全性更高的 AMR；广泛用于仓储、物流和工厂。

The product is equipped with an independently developed LX-MRDVS system, which can accurately identify obstacles and bypass obstacles. AMR with higher safety is widely used in warehousing, logistics and factories.

HPRT DA067D HYBIRD

蝗虫 —— 食物转化机
FROM LOCUST TO NUTRITION

作者：林锦毅 周招坤 麻俊宇
机构：厦门汉印电子技术有限公司
国家：中国
组别：产业组

AUTHOR : Lin Jinyi, Zhou Zhaokun, Ma Junyu
UNIT : Xiamen Hanyin Electronic Technology Co., Ltd.
COUNTRY : China
GROUP : Product Group

作者：汪启韬 郑小涵 陈文静 谢晓文
国家：中国
组别：概念组

AUTHOR : Wang Qitao, Zheng Xiaohan, Chen Wenjing, Xie Xiaowen
COUNTRY : China
GROUP : Concept Group

向定制化和智能制造的数码印花机升级，创造能够用兼具高性价比、高收益、高附加值的数码印花产品解决方案。

Upgrade to customized and intelligently manufactured digital printing machines to create solutions for digital printing products that are cost-effective, profitable, and high value-added.

它是一个诱捕蝗虫并将其粉碎成蛋白质的机器，通过简单的原理，解决非洲贫困地区蝗灾严重、粮食短缺问题。

It is a machine that traps locusts and crushes them into protein, using simple principles to solve locust infestations and food shortages in poor parts of Africa.

巧乐送机器人
PUDU SWIFTBOT

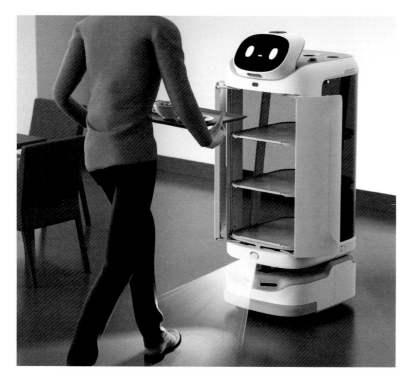

作者：陈鹏 严小龙 杨璐雅 黄思敏
机构：深圳市普渡科技有限公司
国家：中国
组别：产业组

AUTHOR : Chen Peng, Yan Xiaolong, Yang Luya, Huang Simin
UNIT : Shenzhen PUDU Technology Co., Ltd.
COUNTRY : China
GROUP : Product Group

PUDU SWIFTBOT 是基于双激光雷达底盘、全封闭舱体和创新式交互系统组成的全场景配送机器人，应用于餐厅、商场等场景。

The PUDU SWIFTBOT represents a new generation of all-round delivery robots, based on rear dual laser lidar chassis, fully enclosed delivery compartments, and innovative interactive system modules. The robot is suitable for use in indoor venues such as restaurants, and stores.

OMRON DIGITAL THERMOMETER MC-6800B & OMRON CONNECT

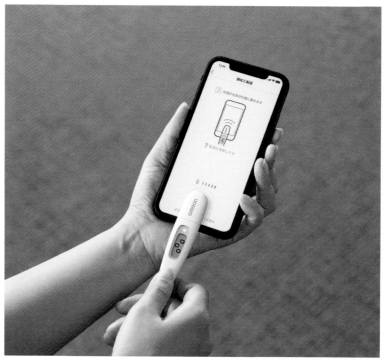

作者：Kosuke Inoue
机构：OMRON HEALTHCARE Co., Ltd.
国家：日本
组别：产业组

AUTHOR : Kosuke Inoue
UNIT : OMRON HEALTHCARE Co., Ltd.
COUNTRY : Japan
GROUP : Product Group

家用数字温度计，采用声波通信技术，仅需15秒即可提供预测温度读数。只需将该设备放在智能手机上，即可传输测量数据。

Home-use digital thermometer that uses sound wave communication technology to deliver predictive temperature readings in just 15 seconds. Simply placing the unit on a smartphone transfers measurement data.

全场景智能骨科手术机器人
FULL-SCENE ORTHOPEDIC SURGERY ROBOT

地震救援机器人
EARTHQUAKE RESCUE ROBOT

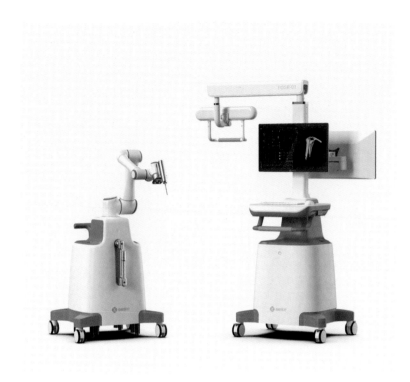

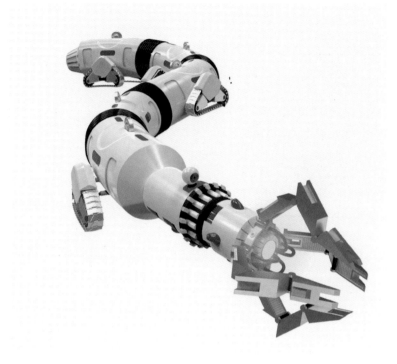

作者：李文倩 彭程 张帅 高丰伟 沈浩
机构：上海由格医疗技术有限公司
国家：中国
组别：产业组

AUTHOR : Li Wenqian, Peng Cheng, Zhang Shuai, Gao Fengwei, Shen Hao
UNIT : Shanghai Youge Medical Technology Co., Ltd.
COUNTRY : China
GROUP : Product Group

作者：王皓辰
国家：中国
组别：概念组

AUTHOR : Wang Haochen
COUNTRY : China
GROUP : Concept Group

一款用于骨科外科手术的微创机器人，能为使用者提供精准定位、微创断骨复位、图像动态匹配和实时可视操作。

A minimally invasive robot for orthopedic surgery can provide users with precise positioning, minimally invasive fracture reduction, image dynamic matching and real-time visual operation.

地震救援蛇形机器人是一款快速进入地震灾区实施救援的机器人，能够极大提高救援效率。

The snake shaped robot for earthquake rescue is a robot that quickly enters the earthquake stricken areas to implement rescue, which can greatly increase the rescue efficiency.

海上救援无人机
RESCUE DRONES AT SEA

作者：曾瑞宸 汪烨 周锦涛 卢鑫源
国家：中国
组别：概念组

AUTHOR : Zeng Ruichen, Wang Ye, Zhou Jintao, Lu Xinyuan
COUNTRY : China
GROUP : Concept Group

具有海上救援功能的无人机设计，概念是通过无人机便携式和快速救援的优点，第一时间进行救援。

The concept of UAV design with sea rescue function is to rescue the first time through the advantages of UAV portable and rapid rescue.

架空输电线路巡检作业机器人
A POWER LINE INSPECTION AND OPERATION ROBOT

作者：陈铭浩 连依晴 汪晗 徐光耀
机构：中国科学院自动化研究所
国家：中国
组别：产业组

AUTHOR : Chen Minghao, Lian Yiqing, Wang Han, Xu Guangyao
UNIT : Institute of Automation, Chinese Academy of Sciences
COUNTRY : China
GROUP : Product Group

架空输电线路巡检作业机器人是一款基于被动柔顺越障方式解决电力巡检与维修问题的作业机器人。

Based on passive compliance obstacle avoidance, power transmission line inspection robot is a kind of robot, which solves the problem of power line inspection and maintenance.

临港 T2 线中运量自动驾驶氢动力数字轨道胶轮电车
T2 AUTONOMOUS DRIVING HYDROGEN ENERGY DRT TRAM

大马力无人驾驶电动拖拉机
HIGH HORSEPOWER DRIVERLESS ELECTRIC TRACTOR

作者：李鲁川 王海涛 李笑帅 崔周森 李付姝玉
机构：中车南京浦镇车辆有限公司 / 脉珈特（上海）设计咨询有限公司
国家：中国
组别：产业组

AUTHOR : Li Luchuan, Wang Haitao, Li Xiaoshuai, Cui Zhousen, Lifu Shuyu
UNIT : CRRC Nanjing Puzhen Co., Ltd. / mattDESIGN (Shanghai) Co.,Ltd.
COUNTRY : China
GROUP : Product Group

作者：毕志强 王启洲 陆在旺 李嘉男 崔晨昊
机构：北京国科廒科技有限公司
国家：中国
组别：概念组

AUTHOR : Bi Zhiqiang, Wang Qizhou, Lu Zaiwang, Li Jianan, Cui Chenhao
UNIT : Beijing guokelin Technology Co., Ltd.
COUNTRY : China
GROUP : Concept Group

产品采用全新流线型外观，司机和乘客体验大幅提升。采用氢燃料电池混合动力、虚拟导向技术、全轮转向等新技术。

New streamlined exterior design, improves the driver and passenger experience. Use new technologies such as hydrogen fuel cell hybrid power, virtual guidance technology, and all-wheel steering.

基于仿生学设计的大马力电动智能农业机械装备。以黑土地课题中的"智能农机装备"为主，通过卫星通信、区域网络进行无人驾驶操控。能源以锂电池模组构成车辆动力，当地的风光发电系统构成储能、换电系统。车辆携带多种传感器，实时记录数据并反馈给数据中心。该产品可实现黑土地的科技农耕，助力农业数字化发展。

This is a high-powered, electric, intelligent agricultural machinery and equipment designed based on biomimicry principles. It primarily focuses on the "intelligent agricultural machinery and equipment" aspect within the context of black soil land. It operates autonomously through satellite communication and regional networks. The vehicle's power source consists of lithium battery modules, while the local wind and solar power systems serve as energy storage and exchange systems. The vehicle is equipped with various sensors that continuously record data and send it to a central data center. This technology-driven agricultural solution contributes to the digitization and advancement of agriculture in black soil regions.

长安
UNI-V

三一背负式 AGV 机器人
SANY AUTOMATED GUIDED VEHICLE

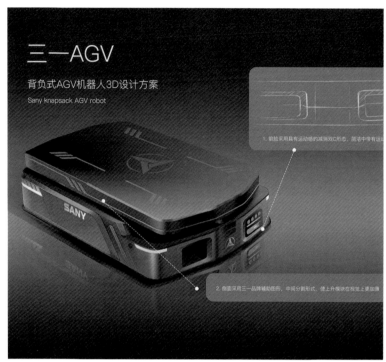

作者：姜楠 周进 王泽晨 雷莉
机构：重庆长安汽车股份有限公司
国家：中国
组别：产业组

AUTHOR : Jiang Nan, Zhou Jin, Wang Zechen, Lei Li
UNIT : Chongqing Changan Automobile Company Limited
COUNTRY : China
GROUP : Product Group

作者：王金磊 杨金龙 林森 辛策 张瑞
机构：三一集团有限公司
国家：中国
组别：产业组

AUTHOR : Wang Jinlei, Yang Jinlong, Lin Sen, Xin Ce, Zhang Rui
UNIT : SANY Group Co.,Ltd.
COUNTRY : China
GROUP : Product Group

长安 UNI-V 是一款面向最年轻用户群体，为他们带来超动态的生活方式与简洁品位的轿跑车。

Changan UNI-V is a coupe aimed at the youngest user group, bringing them an ultra-dynamic lifestyle and simple taste.

三一背负式 AGV 机器人是为工厂、港口及仓储自动化领域等封闭区域作业的自动化、智能化的运输机械，具有自主搬运、自动导航、自动避障、自动充电等功能，在系统的调度下，可以组成一整套完整的搬运自动化系统解决方案。

The SANY Knapsack AGV robot is an automated and intelligent transport machine designed for operations in enclosed areas such as factories, ports, and warehouses in the field of automation. It possesses functions like autonomous carrying, automatic navigation, obstacle avoidance, and self-charging. When coordinated within a system, it can constitute a complete set of material handling automation system solutions.

台式测序实验室
BENCHTOP SEQUENCING LAB

OMRON COMPRESSOR NEBULIZER NE-C28P(NE-C105)

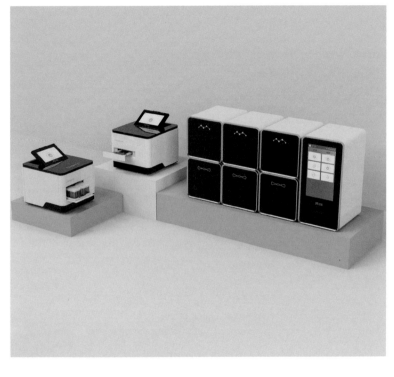

作者：魏诗又 史兴 张引 李开金 喻肖
机构：深圳市华大智造科技股份有限公司
国家：中国
组别：产业组

AUTHOR : Wei Shiyou, Shi Xing, Zhang Yin, Li Kaijin, Yu Xiao
UNIT : MGI Tech Co.,Ltd.
COUNTRY : China
GROUP : Product Group

机构：OMRON HEALTHCARE Co., Ltd.
国家：日本
组别：产业组

UNIT : OMRON HEALTHCARE Co., Ltd.
COUNTRY : Japan
GROUP : Product Group

台式测序实验室是致力于提供轻巧易用可定制组合的桌面测序解决方案的产品。

Benchtop sequencing lab is a product dedicated to providing lightweight, easy-to-use, customizable portfolio desktop sequencing solutions.

这是一款用于哮喘吸入治疗的雾化器。在治疗过程中观察用户的不便是提高易用性的一个关键因素。凭借满足医疗设备标准的高性能水平，该设备通过采用易于操作的药瓶、重新设计的面罩尺寸／形状以及易于使用的气管和开关来提高可用性。

Nebulizer for asthma inhalation therapy. Observing user inconvenience during treatment was a key factor in enhancing ease of use. With a high performance level satisfying medical equipment standards, the unit's improved usability is achieved with an easy-to-handle medication bottle, redesigned mask size/shape, and easy-access air tubing and switches.

小魔驼 2.0
XIAOMOTUO2.0

作者：黄芃超 李忻炳 刘天航
机构：毫末智行科技有限公司 / 洛可可创新设计集团
国家：中国
组别：产业组

AUTHOR : Huang Pengchao, Li Xinbing, Liu Tianhang
UNIT : HAOMO.AI / Rococo Innovative Design (Shenzhen) Co., Ltd.
COUNTRY : China
GROUP : Product Group

小魔驼2.0为末端配送提供了安全、可靠、高效的无人自动驾驶解决方案，作为全球首款十万元级别的无人配送物流车，打破了末端物流自动配送车的成本限制。

With safe, reliable and efficient unmanned driving, Xiaomotuo2.0 is the world's first unmanned delivery robot worth 100,000 Yuan, it leads in cost reduction of unmanned delivery robot for terminal logistics.

VENTILAB IG

作者：BEN KOOK
国家：以色列
组别：产业组

AUTHOR : BEN KOOK
COUNTRY : Israel
GROUP : Product Group

VentiLab IG 产品是一款智能机械通气辅助设备，旨在通过使用正在申请专利的传感器反馈系统，将手持式呼吸机转变为自动响应式通气系统。该呼吸机是在本地制造和采购的，为管理中间通气患者提供了一种低成本、简单的解决方案。

The VentiLab IG product is a Smart Mechanical Ventilation Aid designed to transform a hand held ventilator, into an automatic responsive ventilation system, using a patent pending sensors feedback system. The ventilator is locally manufactured and sourced, and offers a solution cost-effective simple solution for managing intermediate ventilated patients.

野生动物保护相机
WILDLIFE CAMERA

作者：王哲昀 郭李辰 应亮
机构：杭州海康威视数字技术股份有限公司
国家：中国
组别：产业组

AUTHOR : Wang Zheyun, Guo Lichen, Ying Liang
UNIT : Hangzhou Hikvision Digital Technology Co.,Ltd.
COUNTRY : China
GROUP : Product Group

HIKVISION 野生动物相机是一款通过智能 AI 算法，便携灵活布控，以多台组网的方式解决野生动物保护问题的智能产品。

The HIKVISION Wildlife Camera is an intelligent device that employs AI algorithms, portable and flexible deployment, and multi-device networking to address wildlife protection challenges.

梭
SHUTTLE

作者：尹文昊
机构：北京服装学院
国家：中国
组别：概念组

AUTHOR : Yin Wenhao
UNIT : Beijing Institute of Fashion Technology
COUNTRY : China
GROUP : Concept Group

SHUTTLE 是一辆个人载具，为城市交通通勤效率、人的出行效率提供新的思考。

SHUTTLE is a single-person vehicle that provides new thinking for urban traffic commuting efficiency and human travel efficiency.

JYACHT

作者： Umesh Gajbhiye
机构： VIP Industries Ltd.
国家： 印度
组别： 概念组

AUTHOR : Umesh Gajbhiye
UNIT : VIP Industries Ltd.
COUNTRY : India
GROUP : Concept Group

JARVIS 是一艘面向2040年的超级游艇，设计用于航行到海洋的任何角落。这艘游艇由聚变电弧反应堆提供动力，这将有助于最大限度地减少碳足迹。它可用于商务、休闲和探险目的。

JARVIS is a super yacht for 2040, designed to voyage to any corner of the Ocean. It is powered by a Fusion Arc reactor, which will help in reducing the carbon footprints to a maximum level. It can be used for business, leisure, and expedition purposes.

米迦勒医疗生产运输车辆
MICHAEL MEDICAL PRODUCTION TRANSPORT VEHICLE

作者： 付佳文 韩挺 祖悦敏
机构： 上海交通大学设计学院
国家： 中国
组别： 概念组

AUTHOR : Fu Jiawen, Han Ting, Zu Yuemin
UNIT : Shanghai Jiaotong University School of Design
COUNTRY : China
GROUP : Concept Group

将生产过程中的不同生产步骤，转移到同一载体上进行生产。生产和运输一体化的设计会大大增加运输生产效率。

Transfer different production steps in the production process to the same carrier for production. The integrated design of production and transportation will greatly increase the efficiency of transportation production.

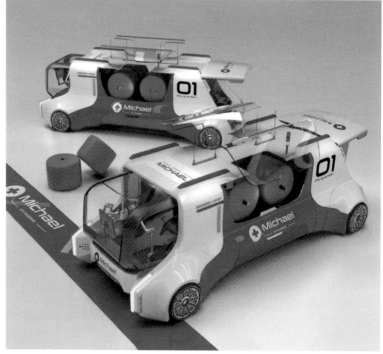

SPREAD KIT

ASH BUNCE

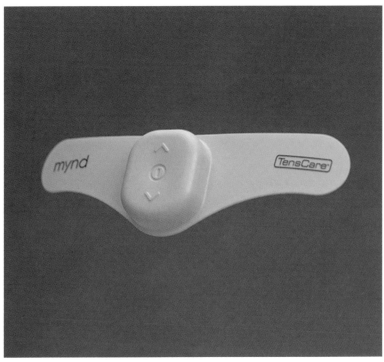

作者：赵浩胜 金勇锡 崔俊日 金艺恩
机构：庆元艺术大学
国家：韩国
组别：概念组

AUTHOR : Hoseung Cho, Yongseok Kim, Junil Choi, Yeeun Kim
UNIT : Kyeowon University of the Arts
COUNTRY : South Korea
GROUP : Concept Group

作者：Ash Bunce, Han Wei Lim, Neil Wright, Kacper Krajewski
机构：TensCare Ltd.
国家：英国
组别：产业组

AUTHOR : Ash Bunce, Han Wei Lim, Neil Wright, Kacper Krajewski
UNIT : TensCare Ltd.
COUNTRY : UK
GROUP : Product Group

涂抹工具是为建筑工程中，从事在墙壁或天花板、地面上涂抹泥土、灰、水泥等职业的人提供最优化的高级装修工具，是帮助其在装修施工中轻松工作的产品。

The applicator tool is an optimised premium finishing tool for people engaged in the occupation of applying clay, plaster and cement to walls or ceilings and floors in construction work, and is a product to help them work easily in renovation work.

Mynd 是一种可穿戴设备，利用经过临床验证的 TENS 技术来应对慢性偏头痛症状。使用黏性凝胶垫，可完美地将该设备置于眉毛上方。温和的微脉冲被发送到三叉神经的上分支，可预防或缓解慢性偏头痛发作。

Mynd is a wearable device that utilises clinically proven TENS technology to manage chronic migraine symptoms. The device sits perfectly above the eyebrows using adhesive gel pads. Gentle micro-impulses are sent to the upper branch of the trigeminal nerve to either prevent or relieve chronic migraine attacks.

机场搭档
AIRPORT PARTNER

作者：卢飞龙
国家：中国
组别：概念组

AUTHOR : Lu Feilong
COUNTRY : China
GROUP : Concept Group

机场搭档通过登机牌信息自动导航不仅可以代步，还可以托运行李以及载人，方便旅客短途转行，是机场好帮手。

Through the automatic navigation of boarding pass information, the airport partner can not only travel, but also check luggage and carry passengers, which is convenient for passengers to change their routes for short distances, and is a good helper for the airport

MANDI

作者：Umesh Gajbhiye, Sachin Sing Tensing, Nikitha S
国家：印度
组别：概念组

AUTHOR : Umesh Gajbhiye, Sachin Sing Tensing, Nikitha S
COUNTRY : India
GROUP : Concept Group

我们开发的设计是一款用于泛印度地区销售水果和蔬菜的自动售货车。该设计的重点是打造一款卫生、模块化和条理分明的购物车，方便买家和卖家使用。

The design that we developed is a vending cart for pan India for selling fruits and vegetables. The design focuses on creating a hygienic, modular, and organized cart that is easily accessible for the buyer and the vendor.

THE TYRE COLLECTIVE

儿童无针注射器
NEEDLE-FREE INJECTOR FOR KIDS

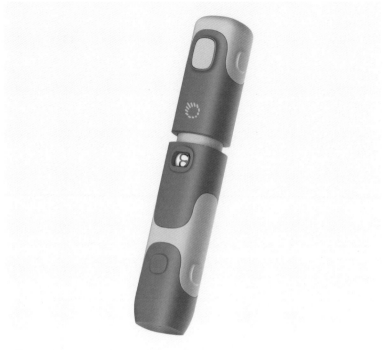

作者：Hanson Cheng, Siobhan Anderson, Hugo Richardson
机构：The Tyre Collective
国家：英国
组别：概念组

AUTHOR : Hanson Cheng, Siobhan Anderson, Hugo Richardson
UNIT : The Tyre Collective
COUNTRY : UK
GROUP : Concept Group

作者：陈苏宁
机构：北京快舒尔医疗技术有限公司
国家：中国
组别：产业组

AUTHOR : Chen Suning
UNIT : Beijing QS Medical Technology Co.,Ltd.
COUNTRY : China
GROUP : Product Group

The Tyre Collective 正在构建解决方案，以获取轮胎磨损情况。轮胎磨损是海洋中第二大微塑料污染物，也是空气污染的一个主要因素。

The Tyre Collective is building solutions to capture tyre wear, the second-largest microplastic pollutant in our oceans and a major contributor to air pollution.

这是一款为儿童设计的无针注射器，消除儿童对注射的恐惧，提高药物的利用度并避免因注射产生的皮下增生。

The needle-free injector designed for kids aims to eliminate their fear of needles, while minimizing such piercing pain, while the localized hardening of soft tissue can be avoided.

新华医疗智慧化 CSSD 整体解决方案
THE OVERALL PLAN OF SHINVA INTELLIGENT DISINFECTION SUPPLY CENTER

作者：王泽坤 韩建康 李现刚 朱书建 任义凯
机构：新华医疗器械股份有限公司
国家：中国
组别：产业组

AUTHOR : Wang Zekun, Han Jiankang, Li Xiangang, Zhu Shujian, Ren Yikai
UNIT : SHINVA
COUNTRY : China
GROUP : Product Group

CSSD 整体解决方案通过物流机器人与设备联动，降低医院消毒供应室由人工操作所导致的感染风险，节省各项资源。

CSSD's overall solution reduces the infection risk caused by manual operations in the hospital disinfection supply room through the linkage of logistics robots and equipment, saving various resources.

小红花微针疫苗
A LITTLE RED FLOWER

作者：缪景怡 邹沍
国家：中国
组别：概念组

AUTHOR : Miao Jingyi, Zou Hu
COUNTRY : China
GROUP : Concept Group

本产品是一款几乎无痛的微针疫苗套组，本产品消除儿童恐针心理，用更符合儿童心理特点的方式治疗疾病。

This product is an almost painless microneedle vaccine set. This product eliminates children's needle phobia and treats diseases in a more friendly and more consistent way with children's psychological characteristics.

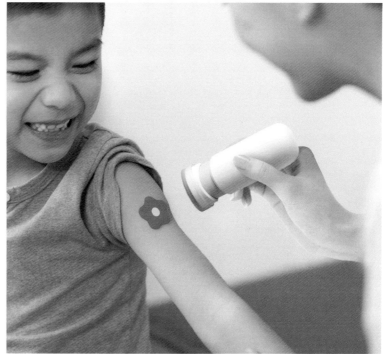

幻肢
V-HANDER

DNR AUTO ANGLE

作者：王翌诚
机构：浙江工业大学设计与建筑学院
国家：中国
组别：概念组

AUTHOR : Wang Yicheng
UNIT : School of Design and Architecture, Zhejiang University of Technology
COUNTRY : China
GROUP : Concept Group

作者：Ahmet brahim POLAT, Aamish UMAR , Kanber SEDEF
机构：DENER MAKNA SAN. VE TC. LTD.ŞT.
国家：土耳其
组别：概念组

AUTHOR : Ahmet brahim POLAT, Aamish UMAR , Kanber SEDEF
UNIT : DENER MAKNA SAN. VE TC. LTD.ŞT.
COUNTRY : Turkey
GROUP : Concept Group

"幻肢"是一款基于眼动控制和 AR 技术来增强患者主动康复意识的多自由度上肢康复外骨骼。

"V-Hander" is a multi-degree-of-freedom upper limb rehabilitation exoskeleton based on eye-movement control and AR display technology to enhance patients' active rehabilitation consciousness.

这是一种利用图像处理技术，借助激光和摄像机设计的非接触式、便携式角度测量装置。

It is a non-contact and portable angle measuring device designed using image processing technology with the help of laser and camera.

C 型移动式 X 光射线机
C-ARM X-RAY

作者 : 张培培 滕轩 陈泽 卢恒
机构 : 南京佗道医疗科技有限公司
国家 : 中国
组别 : 产业组

AUTHOR : Zhang Peipei, Teng Xuan, Chen Ze, Lu Heng.
UNIT : Nanjing Tuodao Medical Technology Co., Ltd.
COUNTRY : China
GROUP : Product Group

移动式平板 C 臂是一款为骨科以及疼痛微创介入透视成像的医疗设备。

The mobile flat-panel C-arm is a medical device for orthopedic and pain minimally invasive interventional fluoroscopic imaging.

V3 呼吸机
V3 VENTILATOR

作者 : 钟琳琳 匡思能 周妙雨
机构 : 深圳市科曼医疗设备有限公司
国家 : 中国
组别 : 产业组

AUTHOR : Zhong Linlin, Kuang Sineng, Zhou Miaoyu
UNIT : Shenzhen Comen Medical Instruments Co., Ltd.
COUNTRY : China
GROUP : Product Group

V3呼吸机是一款多功能应用呼吸机，能够提供从院内急救转运到危重症治疗全程的呼吸治疗解决方案。

V3 ventilator is a multi-functional application ventilator that can provide respiratory treatment solutions from in-hospital emergency transport to critical care treatment.

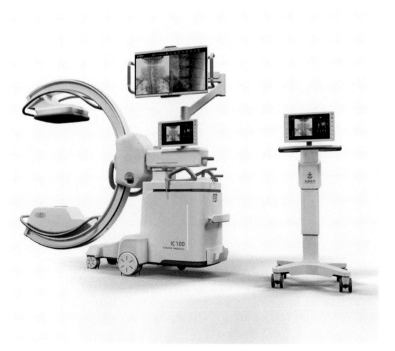

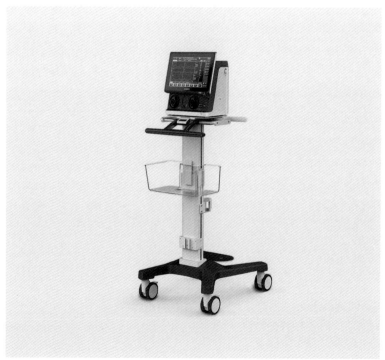

鱼跃电子体温计 YT-3
YUWELL ELECTRONIC THERMOMETER YT-3

作者：施逸琪 杨天照 华昊 邱博 高昱明
机构：江苏鱼跃医疗设备股份有限公司
国家：中国
组别：产业组

AUTHOR : Shi Yiqi, Yang Tianzhao, Hua Hao, Qiu Bo, Gao Yuming
UNIT : JIANGSU YUWELL MEDICAL EQUIPMENT & SUPPLY Co., Ltd.
COUNTRY : China
GROUP : Product Group

YT-3是一款快捷方便且便于收纳的家用电子体温计，用于快速测量用户的前额体温。

YT-3 is a fast,convenient and easy-to-store household electronic thermometer for quickly measuring the user's forehead temperature.

鱼跃快速检测系列卡壳
YUWELL RAPID SELF-TEST KITS

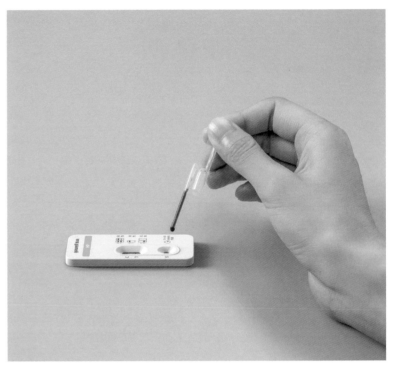

作者：郑怡珺 华昊 施逸琪 赵扬 杨天照
机构：江苏鱼跃医疗设备股份有限公司
国家：中国
组别：产业组

AUTHOR : Zheng Yijun, Hua Hao, Shi Yiqi, Zhao Yang, Yang Tianzhao
UNIT : JIANGSU YUWELL MEDICAL EQUIPMENT & SUPPLY Co., Ltd.
COUNTRY : China
GROUP : Product Group

鱼跃快速检测卡壳是一系列轻便小巧、易于判读的自检产品，覆盖包括 HIV, 幽门螺旋杆菌等在内的多个检测项目。

Yuwell rapid self-test kits are a series of slim and easy-to-use self-test products. The series cover multiple test items including HIV ,gastric Helicobacter pylori antigen, etc.

鱼跃臂式一体血压计 630CR
YUWELL UPPER ARM BLOOD PRESSURE MONITOR 630CR

作者：杨天照 施逸琪 高昱明 卞程莹 华昊
机构：江苏鱼跃医疗设备股份有限公司
国家：中国
组别：产业组

AUTHOR : Yang Tianzhao, Shi Yiqi, Gao Yuming, Bian Chengying, Hua Hao
UNIT : JIANGSU YUWELL MEDICAL EQUIPMENT & SUPPLY Co., Ltd.
COUNTRY : China
GROUP : Product Group

630CR 是一款一体式血压计，为商务差旅人士打造极致便携、测量精准、智能互联的体验。

The 630CR is a smart and compact BPM on-the-go for travelers. It provides extremely convenient, accurate, and intelligent blood pressure monitoring experience.

鱼跃安耐糖 CT3 持续葡萄糖监测系统
YUWELL CT3 CONTINUOUS GLUCOSE MONITORING SYSTEM

作者：孙博珍 华昊 史小雅 施逸琪 杨天照
机构：江苏鱼跃医疗设备股份有限公司
国家：中国
组别：产业组

AUTHOR : Sun Bozhen, Hua Hao, Shi Xiaoya, Shi Yiqi, Yang Tianzhao
UNIT : JIANGSU YUWELL MEDICAL EQUIPMENT & SUPPLY Co., Ltd.
COUNTRY : China
GROUP : Product Group

鱼跃安耐糖 CT3是一款为血糖异常用户（如糖尿病人）提供全天24小时不间断血糖监测的软硬件解决方案。

Yuwell Anytime CT3 CGM is a solution providing continuous blood glucose monitoring 24 hours a day for users with blood glucose related issues, such as diabetics.

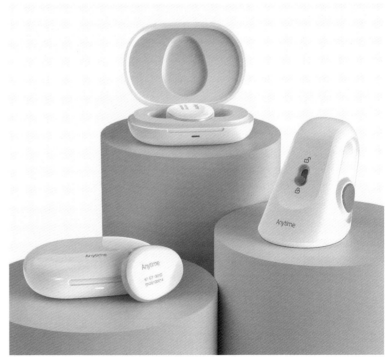

华诺康4K 内窥镜系统
THE HEALNOC 4K ENDOSCOPY SYSTEM

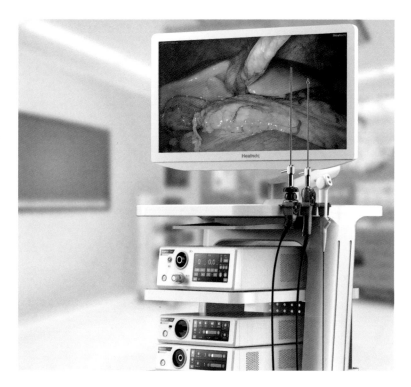

作者：魏亚军 陈力 卢强博 王胜浩 靳华玲
机构：浙江华诺康科技有限公司 / 浙江大华技术股份有限公司
国家：中国
组别：产业组

AUTHOR : Wei Yajun, Chen Li, Lu Qiangbo, Wang Shenghao, Jin Hualing
UNIT : Zhejiang Healnoc Technology Co.,Ltd. / Dahua Technology Co., Ltd.
COUNTRY : China
GROUP : Product Group

自主研发的4K 内窥镜系统，具有出色的精度和光学性能，能更快更准确地识别病变，实现高端医疗产品的国产化。

The self-developed 4K endoscope system has excellent precision and optical performance, can identify lesions faster and more accurately, and realize the localization of high-end medical products.

悠然210下肢坐卧式康复训练器
URA210 LOWER LIMB REHABILITATION TRAINER

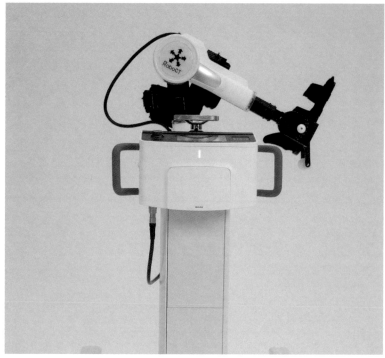

作者：颜海 王天 赵晴宇 李聪 叶广兴
机构：杭州程天科技发展有限公司
国家：中国
组别：产业组

AUTHOR : Yan Hai, Wang Tian, Zhao Qingyu, Li Cong, Ye Guangxing
UNIT : Hangzhou RoboCT Technology Development Co.,Ltd.
COUNTRY : China
GROUP : Product Group

悠然210下肢坐卧式康复训练器是一款提供主被动训练，解决髋膝关节术后、偏瘫下肢康复问题的智能医疗产品。

URA210 lower limb rehabilitation trainer is an intelligent medical product that provides active and passive training to solve the problems of lower limb rehabilitation after hip and knee joint surgery and hemiplegia.

智能动力防水小腿假肢
WATERPROOF SMART POWERED TRANSTIBIAL PROSTHESIS P105

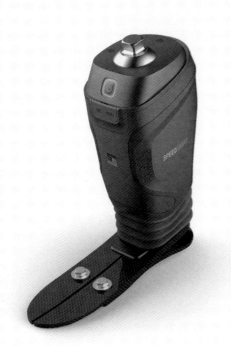

X66 麻醉工作站
X66 ANESTHESIA SYSTEM

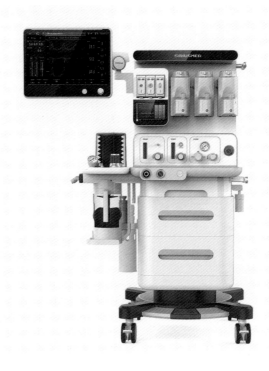

作者：罗嶷 汤晓杭
机构：北京工道风行智能技术有限公司 / 深圳市摩迪赛产品设计管理有限公司
国家：中国
组别：产业组

AUTHOR : Luo Yi, Tang Xiaohang
UNIT : Beijing Speedsmart Technology Co., Ltd. / Shenzhen Modesign Management Co.,Ltd.
COUNTRY : China
GROUP : Product Group

通过不断的数据积累及算法优化解决普通假肢无法做到的步态仿真问题，帮助障碍人士能够正常生活。

Through continuous data accumulation and algorithm optimization, it can solve the gait simulation that ordinary prosthetics cannot do, and autonomously identify human motion intentions. IP65 waterproof design brings great convenience to users.

作者：赵东阁 王延庆 陈定亮 刘国新 魏璟涛
机构：北京思瑞德医疗器械有限公司 / 北京智加问道科技有限公司
国家：中国
组别：产业组

AUTHOR : Zhao Dongge, Wang Yanqing, Chen Dingliang, Liu Guoxin, Wei Jingtao
UNIT : Beijing Siriusmed Medical Equipment Co., Ltd. / ZCO Design Co., Ltd.
COUNTRY : China
GROUP : Product Group

它是给患者提供吸入麻醉并在手术中全面监测患者的生命体征参数，高度集成化、高度智能化的医疗设备。

It is a highly integrated and intelligent medical equipment that can provide patients with inhalation anesthesia and comprehensively monitor the vital signs parameters of patients.

Typdont+

作者： Luke Goh Xu Jie, Loh Yue Xuan Chantel, Hugo Jean Guillaume
机构： National University of Singapore (NUS)
国家： 新加坡
组别： 产业组

AUTHOR : Luke Goh Xu Jie, Loh Yue Xuan Chantel, Hugo Jean Guillaume
UNIT : National University of Singapore (NUS)
COUNTRY : Singapore
GROUP : Product Group

Typodont+ 是世界上第一款微型螺钉模拟器，能够模拟各种人类颌骨，并确定微型螺钉是否与牙根接触。它允许牙医在真正的患者面前练习将微型螺钉（一种正畸固定锚固装置）插入模拟器模型中。

Typodont+ is the world's first mini screw simulator capable of simulating various human jawbones and determining whether a mini screw has made contact with a tooth root. It allows dentists to practice inserting mini-screws, a type of orthodontic temporary anchorage device, into the simulator model before an actual patient.

下肢康复机器人
LOWER LIMB REHABILITATION ROBOT

作者： 赵智峰 芮达志 唐聪智 陈亚新 李得阳
机构： 上海璟和技创机器人有限公司 / 苏州柒整合设计有限公司
国家： 中国
组别： 产业组

AUTHOR : Zhao Zhifeng, Rui Dazhi, Tang Congzhi, Chen Yaxin, Li Deyang
UNIT : Shanghai Jinghe Technology Robotics Co., Ltd. / Suzhou CHY Integration Design Co., Ltd.
COUNTRY : China
GROUP : Product Group

下肢康复机器人简化传统康复治疗中"一对一"繁重的治疗过程，实现有针对性的、重复性的训练。

Lower limb rehabilitation robot can simplify the onerous treatment process of "one to one" in traditional rehabilitation therapy, and achieve targeted and repetitive training.

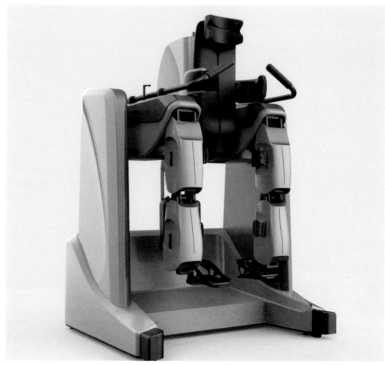

ARTEMIS 389

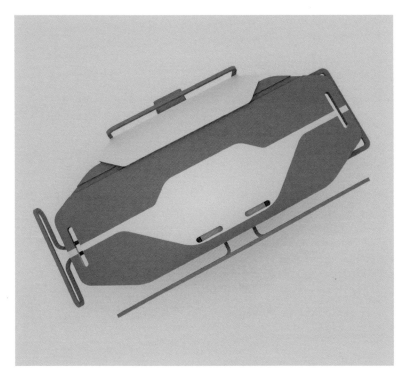

作者：Anuja Tripathi
国家：印度
组别：概念组

AUTHOR : Anuja Tripathi
COUNTRY : India
GROUP : Concept Group

Artemis 389是一种患者转运设备，在医院内进行患者转运时，可为患者和医护人员提供轻松、无痛苦的体验。它使患者转运变得更简单、更顺畅、更省时。

Artemis 389 is a patient transfer equipment that provides an effortless and pain-free experience to the patients and the healthcare workers while performing patient transfers inside hospitals. It makes patient transfers simpler, smoother and less time-consuming.

BOB 儿童康复平衡推举仪
BOB CHILDREN'S REHABILITATION BALANCE PRESS

作者：任紫涵 王龙龙
国家：中国
组别：概念组

AUTHOR : Ren Zihan, Wang Longlong
COUNTRY : China
GROUP : Concept Group

这是一款通过模块化的磁吸推手和可调控的平衡板结合的方式解决儿童上肢功能障碍问题的产品。

This is a product that solves the problem of children's upper limb dysfunction through the combination of modular magnetic suction push hands and adjustable balance plates.

智慧医疗 ——下肢康复穿戴设备
WITMED — LOWER LIMB REHABILITATION WEARABLE DEVICES

作者：秦桂祥 许佳 胡浩伟
国家：中国
组别：概念组

AUTHOR : Qin Guixiang, Xu Jia, Hu Haowei
COUNTRY : China
GROUP : Concept Group

智慧医疗 - 下肢康复穿戴设备是一款改善康复枯燥，完善训练计划且轻量化的家庭辅助康复产品。

WITMED-Lower Limb Rehabilitation Wearable Device Is a lightweight home-assisted rehabilitation product that improves rehabilitation boredom, perfects training plans and is complete.

磁共振引导肝癌消融智能手术机器人
MRI GUIDED SURGICAL ROBOT DESIGN FOR LIVER CANCER

作者：孙博文 袁少玫 李迪嘉 汤毅 王嘉
机构：北京精准医械科技有限公司 / 北京理工大学设计与艺术学院
国家：中国
组别：产业组

AUTHOR : Sun Bowen, Yuan Shaomei, Li Dijia, Tang Yi, Wang Jia
UNIT : Beijing Precision Medtech Technologies Co.,Ltd. / School of Design & Arts,Beijing Institute of Technology
COUNTRY : China
GROUP : Product Group

国内首创磁共振兼容机械臂和控制台引导穿刺与温控，设计整合优化全域体验，解决肝癌消融的智能手术机器人。

The first MRI-compatible robotic arm and console in Chinese to guide puncture and temperature control, which designed an intelligent surgical robot that integrates and optimates the whole experience to solve liver cancer ablation.

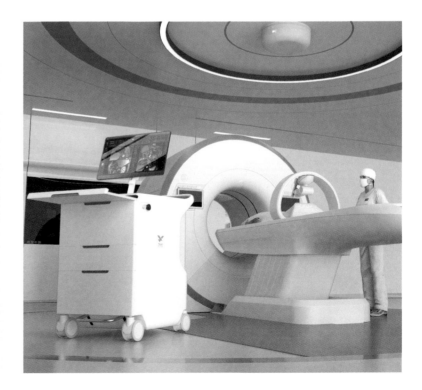

AR 骨科手术导航系统
HOLONAVI LUMBAR PUNCTURE OUTFIT

作者：高凤麟 刘沛桐 刘洋
机构：上海霖晏医疗科技有限公司 / 心冥想健康科技（杭州）有限公司
国家：中国
组别：产业组

AUTHOR：Gao Fenglin, Liu Peitong, Liu Yang
UNIT：Shanghai Linyan Medical Technology Co., Ltd. / Shine Meditation Health Technology (Hangzhou) Co., Ltd.
COUNTRY：China
GROUP：Product Group

使用 NDI 的红外设备协助医生进行脊柱穿刺手术，使医生更准确地找到正确位置，提高手术成功率，降低手术风险。

Medical instruments use NDI infrared equipment to assist doctors in spinal puncture surgery. So that doctors can find the correct position more accurately, improve the success rate of surgery and reduce the risk of surgery.

载人 eVTOL 飞行器（ZG-ONE）
MANNED EVTOL AIRCRAFT（ZG-ONE）

作者：贾思源 李宜恒 班剑锋 黄宇欣 陈邵洁
机构：零重力深圳飞机工业有限公司
国家：中国
组别：产业组

AUTHOR：Jia Siyuan, Li Yiheng, Ban Jianfeng, Huang Yuxin, Chen Shaojie
UNIT：Zero-G Shenzhen Aircraft Industry Co.,Ltd.
COUNTRY：China
GROUP：Product Group

ZG-ONE 是一款通过垂直起降功能来满足城市空中立体交通的场景需求，从而缓解目前城市交通拥堵困境的载人飞行器。

ZG-ONE is a manned aircraft that meets the scene requirements of urban aerial three-dimensional traffic through the vertical take-off and landing function, thereby alleviating the current urban traffic congestion dilemma.

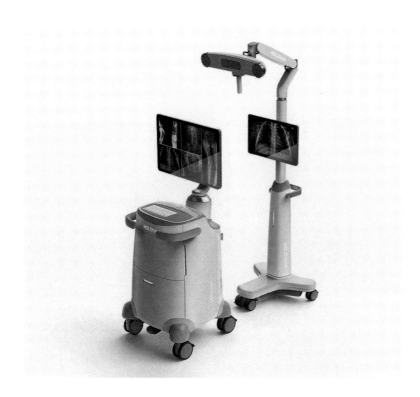

NKB5000 —— 视控一体键盘
NKB5000 KEYBOARD

基于并联旁路动力变速的铰接式山地拖拉机
ARTICULATED MOUNTAIN TRACTOR

作者：刘雅慧 陈力 李玮 杨康 彭皓明
机构：浙江大华技术股份有限公司
国家：中国
组别：产业组

AUTHOR : Liu Yahui, Chen Li, Li Wei, Yang Kang, Peng Haoming
UNIT : Dahua Technology Co., Ltd.
COUNTRY : China
GROUP : Product Group

作者：周宁 罗钦林 李荣 廖建群 唐刚
机构：长沙九十八号工业设计有限公司
国家：中国
组别：产业组

AUTHOR : Zhou Ning, Luo Qinlin, Li Rong, Liao Jianqun, Tang Gang
UNIT : N98Design
COUNTRY : China
GROUP : Product Group

这是一款分体带屏布控键盘，优势在于快速部署，操作便捷，为布控创造极佳且更高效的用户体验。

This is a split control keyboard, which has the advantages of rapid deployment and convenient operation, creating an excellent and more efficient user experience for control.

这是一款采用并联旁路动力变速技术的铰接式山地拖拉机，适合丘陵山区地区使用，提高了产品的动力性和经济性。

This is an articulated mountain tractor with parallel bypass power transmission technology, which is suitable for hilly and mountainous areas and improves the power and economy of the product.

VOGUE HIGHBAY

作者： Sushmita Jaiswal, Sumit Singh, Deep Singh
国家： 印度
组别： 概念组

AUTHOR : Sushmita Jaiswal, Sumit Singh, Deep Singh
COUNTRY : India
GROUP : Concept Group

节能型智能吊灯，采用通用外形，易于维护、维修和安装，采用可持续和环保的方法制造。

Energy efficient smart highbay with universal shape which is made for human ease-ease of maintenance, easy serviceability & easy installation. Made with sustainable & environmental friendly approach.

聋哑人手势智能翻译臂环
INTELLIGENT GESTURE ARMBAND FOR THE DEAF

作者： 滕佳琪 曾芷涵 徐子昂 夏涵飞 厉向东
机构： 浙江大学
国家： 中国
组别： 产业组

AUTHOR : Teng Jiaqi, Zeng Zhihan, Xu Ziang, Xia Hanfei, Li Xiangdong
UNIT : Zhejiang University
COUNTRY : China
GROUP : Product Group

聋哑人手势智能翻译臂环是一款检测实时手势肌电信号并准确转换为语音，从而解决聋哑人正常沟通问题的产品。

Intelligent gesture translation armband for the deaf is a product that detects real-time gesture EMG signals and accurately converts them into speech to solve the problem of normal communication for deaf people.

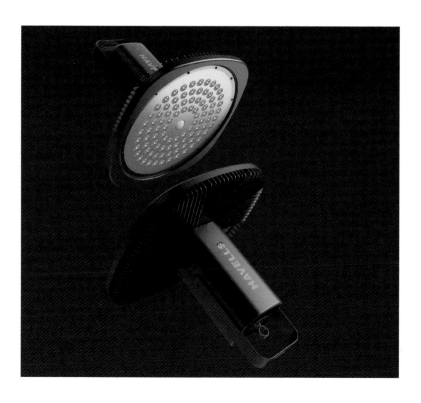

三一电动重卡车头
SANY ELECTRIC HEAVY TRUCK HEAD

极22三维扫描仪
EXTR22 3D SCANNER

作者：朱宏 吕振广 刘海军 陈勇 郭海亮
机构：三一集团有限公司
国家：中国
组别：产业组

作者：盛明圆 茹方军 李冠楠
机构：杭州非白三维科技有限公司
国家：中国
组别：产业组

AUTHOR : Zhu Hong, Lv Zhenguang, Liu Haijun, Chen Yong, Guo Hailiang
UNIT : SANY Group Co.,Ltd.
COUNTRY : China
GROUP : Product Group

AUTHOR : Sheng Mingyuan, Ru Fangjun, Li Guannan
UNIT : FORMBUILDER
COUNTRY : China
GROUP : Product Group

重型卡车的车头，如货车、牵引车、自卸车、搅拌车等。

Used as the head of heavy trucks such as vans, tractors, dump trucks, mixers.

"极22"是一款通过3D机器视觉技术，对物体实现高精度（0.01mm）手持式蓝光3D扫描建模的产品。

"EXTR22"is a handheld device designed to conduct high-precision (0.01 mm) 3D scanning and modelling of objects through machine vision technology.

VT-30

作者：胡华智 李智奕 马春明 陈娜
机构：亿航智能设备（广州）有限公司
国家：中国
组别：产业组

AUTHOR : Hu Huazhi, Li Zhiyi, Ma Chunming, Chen Na
UNIT : EHang
COUNTRY : China
GROUP : Product Group

VT-30是专为城际间空中交通设计的机型，可容纳2人，可为城市居民提供更加省时、高效、低成本的交通解决方案。

Vt-30 is a model specially designed for Intercity air traffic, with a capacity of 2 people. It is expected to provide more time-saving, efficient and low-cost transportation solutions for urban residents.

TO SEE OR NOT TO SEE

作者：Tuncay ince
国家：土耳其
组别：概念组

AUTHOR : Tuncay ince
COUNTRY : Turkey
GROUP : Concept Group

"TO SEE OR NOT TO SEE" 的行李箱设计，可以是不透明的，也可以是透明的，在功能上向使用者呈现艺术和技术。该产品旨在消除人们在旅途中进行强制性或选择性行李检查过程中打开行李时的时间损失和犹豫。

"TO SEE OR NOT TO SEE"luggage design, which can be opaque and transparent, presents art and technology to its users functionally.The product aims to eliminate the loss of time and hesitation in opening luggage during mandatory or optional luggage checks during people's journeys.

HYDROSURV

方圆号 —— 75m 东方美学商务游艇概念设计
FANG YUAN — 75M ORIENTAL BUSINESS YACHT CONCEPT

作者：David Hull
机构：HydroSurv
国家：英国
组别：产业组

AUTHOR : David Hull
UNIT : HydroSurv
COUNTRY : UK
GROUP : Product Group

作者：曹中淏 王萌 李博 吴亮 孙天为
国家：中国
组别：概念组

AUTHOR : Cao Zhonghao, Wang Meng, Li Bo, Wu Liang, Sun Tianwei
COUNTRY : China
GROUP : Concept Group

HydroSurv 是电动和混合动力无人水面舰艇（USV）领域的全球创新者，其长度为 1.6~6 米，用于内陆、近海和海上的数据采集。我们的商业模式普及了 USV 平台的可获得性，因此我们的客户可以大幅削减水文、海洋和环境调查项目的成本和碳强度。

HydroSurv is a global innovator in electric and hybrid Uncrewed Surface Vessels (USVs) from 1.6~6m for inland, nearshore and offshore data acquisition. Our business model democratises the accessibility of USV platforms so our customers can slash the cost and carbon-intensity of hydrographic, oceanographic and environmental survey projects.

方圆号是一艘体现东方文化价值的、集商务与休闲为一体的 75 米概念游艇。

FANG YUAN is a 75m concept yacht that embodies the value of Oriental culture and integrates business and leisure.

未来智能交通巴士
FUTURE INTELLIGENT TRANSPORTATION BUS

飞凡 R7
RISING AUTO R7

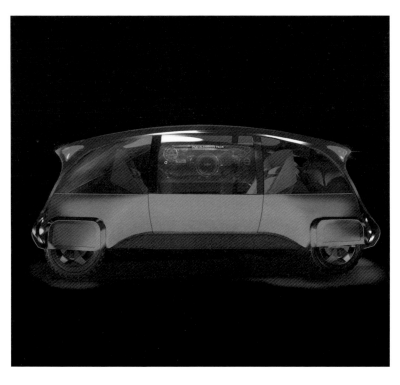

作者：刘卉媛
国家：中国
组别：概念组

AUTHOR : Liu HuiYuan
COUNTRY : China
GROUP : Concept Group

作者：邵景峰 邵长山 齐精文 王睿 黄晴辉
机构：上汽集团 / 上汽设计中心
国家：中国
组别：产业组

AUTHOR : Shao Jingfeng, Shao Changshan, Qi Jingwen, Wang Rui, Huang Qinghui
UNIT : SAIC GROUP / SAIC DESIGN CENTER
COUNTRY : China
GROUP : Product Group

这是一款智能巴士概念设计，以小模块的运行方式，进行非固定式的路线运输，打造快速出门的便捷交通系统。

This is a smart bus concept designed to operate in small modules for non-fixed route transport, creating a convenient transport system for getting out and about quickly.

上汽集团高端纯电汽车品牌 —— 飞凡汽车，上汽设计主导的 R7 是飞凡汽车旗舰车型，车辆定位为中大型 SUV。

SAIC Group has newly established a high-end pure electric automobile brand - Rising-auto. As the flagship model of Feifan automobile, R7 led by SAIC design center is positioned as a medium-sized and large-scale SUV, which adopts the industry-leading advanced intelligent driving technology, reflecting SAIC's strong scientific and technological strength.

可持续清洁公共洗车机器人系统
SUSTAINABLE CLEANING PUBLIC CAR WASH ROBOT SYSTEM

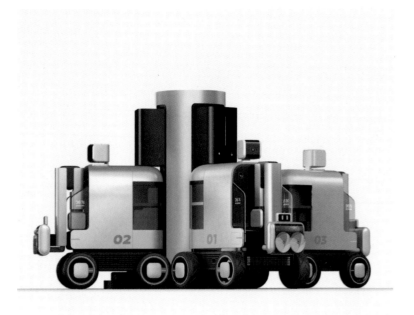

美团 X1共享电单车
MEITUAN X1 SHARING E-SCOOTER

作者 : 刘志强 林树毫
机构 : 广州美术学院
国家 : 中国
组别 : 概念组

AUTHOR : Liu Zhiqiang, Lin Shuhao
UNIT : Guangzhou Academy of Fine Arts
COUNTRY : China
GROUP : Concept Group

公共洗车机器人系统是一款通过可持续清洁方式解决洗车服务水资源浪费和承载量不足问题的公共洗车服务产品。

The public car wash robot system is a public car wash service product that solves the problem of water waste and insufficient carrying capacity in car wash services through sustainable cleaning methods.

作者 : 刘译元 赵猛 章思远 邹慧琳 毛非一
机构 : 美团
国家 : 中国
组别 : 产业组

AUTHOR : Liu Yiyuan, Zhao Meng, Zhang Siyuan, Zou Huilin, Mao Feiyi
UNIT : Meituan
COUNTRY : China
GROUP : Product Group

X1是一款通过优化骑行体验，提升外观设计，丰富安全保护功能，解决传统共享产品粗制滥造问题的共享电单车。

X1 is a new generation of sharing E-scooter which solves the problems of shabby appearance and poor safety performance of traditional products on the market by optimizing the riding experience, improving the industrial design, and enriching the safety protection functions.

818防疫安检通道
ULTRA ENTRANCE GATE

作者：薛李安 陈力 杨易铭 李示明 黄祝秋
机构：浙江华视智检科技股份有限公司 / 浙江大华技术股份有限公司
国家：中国
组别：产业组

AUTHOR : Xue Li'an, Chen Li, Yang Yiming, Li Shiming, Huang Zhuqiu
UNIT : Zhejiang Huajian Technology Co.,Ltd. / Dahua Technology Co., Ltd.
COUNTRY : China
GROUP : Product Group

818防疫安检通道，是一款颠覆传统疫情安检形式的一站式、全自动、无感化、高效率安检设备。

Ultra Entrance Gate is a fully automatic, one-stop senseless security inspection device that overturned the traditional security inspection form.

双差速舵轮驱动重载 AGV
DOUBLE DIFFERENTIAL HELM DRIVES HEAVY DUTY AGV

作者：范育芳 莫光辉 胡徐起 叶丹丹 吕荣平
机构：汉度（杭州）创意设计发展有限公司
国家：中国
组别：产业组

AUTHOR : Fan Yufang, Mo Guanghui, Hu Xuqi, Ye Dandan, Lv Rongping
UNIT : Hando Design
COUNTRY : China
GROUP : Product Group

双差速舵轮驱动 AGV 是一款实现货物和物流装卸与搬运全过程的自动化机械产品。

Double differential helm drives heavy-duty AGV is an automatic mechanical product that realizes the whole process of loading, unloading and handling of goods and logistics.

范德兰德 SBD 机场自助值机系统
VANDERLANDE SBD AIRPORT SELF CHECK-IN SYSTEM

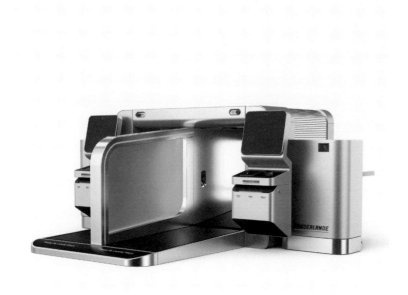

作者：马海豹 朱池 刘昆 顾闻 郑文雄
机构：范德兰德物流自动化系统（上海）有限公司 / 上海木马工业产品设计有限公司
国家：中国
组别：产业组

AUTHOR : Ma Haibao, Zhu Chi, Liu Kun, Gu Wen, Zheng Wenxiong
UNIT : Vanderlande Industries Logistics Automated Systems (Shanghai) Co., Ltd. /
Shanghai Muma Industrial Products Design Co., Ltd.
COUNTRY : China
GROUP : Product Group

范德兰德 SBD 机场自助值机系统，是一款用于乘客自助值机以及自动化行李托运的机场行李处理系统。

Vanderlande SBD Airport Self-Service Check-in is an airport baggage handling system for passenger self-service check-in and automated baggage drop-off.

为载人登火任务服务的子母火星车系统概念设计
MOTHER-CHILD MARS ROVER FOR MANNED FIRE MISSIONS

作者：李自翔
国家：中国
组别：概念组

AUTHOR : Li Zixiang
COUNTRY : China
GROUP : Concept Group

通过设计基于子母协作系统运作的火星车，来满足未来火星探测任务的需求，解决或缓解现有漫游车的局限性。

By designing a Mars rover that operates based on the mother-child collaboration system architecture, it can meet the needs of future Mars exploration missions and solve or alleviate the limitations of existing rovers.

SD-10

作者：Hitoshi Igarashi, Manabu Kawahara, Hideki Kato, Shigeaki Isobe
机构：SEIKO EPSON CORPORATION
国家：日本
组别：产业组

AUTHOR : Hitoshi Igarashi, Manabu Kawahara, Hideki Kato, Shigeaki Isobe
UNIT : SEIKO EPSON CORPORATION
COUNTRY : Japan
GROUP : Product Group

SD-10分光光度计不仅精度高、结构紧凑，而且价格合理，同时还配备了我们内部开发的 MEMS 法布里 - 珀罗腔（Fabry-Perot）可调光学滤波器。这款分光光度计与智能手机或云服务相链接，完美实现色彩匹配数字化，用户在使用时可进行色彩信息集中管理，打印也更为流畅。另外，这款产品体积小巧、方便携带，可随时随地测量色彩。

Highly-accurate, compact, and affordable spectrophotometer features MEMS Fabry-Perot Tunable Filter developed in-house. Use it to digitalize color matching, a conventionally time-consuming part of printing done manually, and link it with smartphone or cloud-service to centrally manage color information for smoother printing. Pocket-sized and portable to measure colors anytime, anywhere.

THE HOMEOWNER'S HYDROPONICS

作者：Nahita Zafimahova
国家：肯尼亚
组别：概念组

AUTHOR : Nahita Zafimahova
COUNTRY : Kenya
GROUP : Concept Group

一家包罗万象的电子商务商店，面向想要开始水培并想自己种植食物的阳台园丁。

The Homeowner's Hydroponics is an all encompassing E-commerce shop for balcony gardeners who want to get started in hydroponics and who want to grow their own food.

数字经济
DIGITAL ECONOMY

支付宝盒 R0
ALIPAY R0

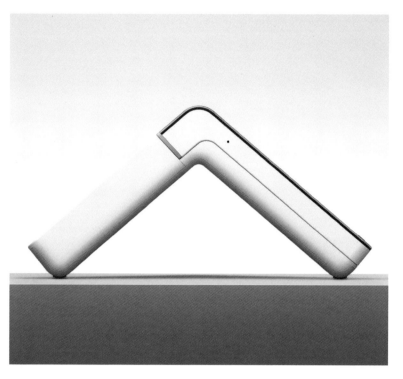

作者: 张铭伟 陈志远
机构: 支付宝
国家: 中国
组别: 产业组

AUTHOR: Zhang Mingwei, Chen Zhiyuan
UNIT: Alipay
COUNTRY: China
GROUP: Product Group

支付宝盒 R0面向小微商户。以前开店购买一系列硬件的费用过高,因此我们设计了这款"一站式"软硬件产品。

Alipay R0 targets the micro-sized merchants, who face difficulties in placing a family of hardware (cash register, POS terminal etc.). We specially designed this all-in-one device and significantly reduced the number of devices needed by them, thereby reducing the overall purchase cost.

微信刷掌服务
WEPALM

作者: 张颖 叶娃 侯锦坤 郭润增 黄家宇
机构: 腾讯科技(深圳)有限公司
国家: 中国
组别: 产业组

AUTHOR: Zhang Ying, Ye Wa, Hou Jinkun, Guo Runzeng, Huang Jiayu
UNIT: Tencent Technology (Shenzhen) Co.,Ltd.
COUNTRY: China
GROUP: Product Group

WePalm 是一套刷掌服务,能适配多种场景使用,透过"一掌通"串联整体服务,让用户享受便捷、安全又卫生的交互体验。

WePalm is a set of products that provides palm recognition services, which can be quickly adapted to a variety of scenarios, and allows users to enjoy a convenient, safe and hygienic interactive experience by linking the overall services through "One Palm".

魔派系列智能终端
MAGIC PAD SMART TERMINAL

作者：黄立清 廖鹭蓉 蔡玉敏
机构：厦门立林科技有限公司
国家：中国
组别：产业组

AUTHOR：Huang Liqing, Liao Lurong, Cai Yumin
UNIT：XIAMEN LEELEN TECHNOLOGY Co., Ltd.
COUNTRY：China
GROUP：Product Group

魔派系列终端产品是全屋智能化的控制终端，同时它也是智能门禁，可以与室外主机连接，守护家人的安全。

Magic Pad smart terminal products are intelligent control terminals for the whole house, and they are also intelligent access control devices that can be connected with outdoor hosts to protect the safety of family members.

WATERFALL-NYC

机构：D'STRICT KOREA, INC.
国家：韩国
组别：产业组

UNIT：D'STRICT KOREA, INC.
COUNTRY：South Korea
GROUP：Product Group

"Waterfall-NYC" 是一个公共媒体艺术产品，利用四个垂直连接到纽约时代广场一座外墙的 LED 屏幕，呈现出一个高达128米的巨大数字瀑布。

"Waterfall-NYC" is a public media art that expresses a huge 128m high digital waterfall using four LED screens vertically connected to the exterior wall of One Times Square in Times Square, New York.

吉利银河车机系统
GEELY GALAXY OS

作者：孙涛 陈思聪 陈江学涯 吴浩 肖婷婷
机构：吉利控股集团 / 亿咖通科技
国家：中国
组别：产业组

AUTHOR : Sun Tao, Chen Sicong, Chen Jiangxueya, Wu Hao, Xiao Tingting
UNIT : GEELY / ECARX
COUNTRY : China
GROUP : Product Group

吉利银河车机系统以自主研发的芯片与 AI 技术串联汽车、手机和家庭，将汽车变为智能空间，用户量超 400 万。

Geely Galaxy OS connects the car, the mobile phone, and the home with its in-house-developed chip and AI technology, turning the car into an intelligent space with over 4 million users.

智己 L7 智能驾舱系统(IMOS)
INFORMATION SYSTEM HMI DESIGN FOR IM L7 (IMOS)

作者：李微萌 关超雄 胡青剑 彭璆 方贻刚
机构：智己汽车 / 智己汽车·软件 & 用户触点团队
国家：中国
组别：产业组

AUTHOR : Li Weimeng, Guan Chaoxiong, Hu Qingjian, Peng Qiu, Fang Yigang
UNIT : IM Motors / IUED
COUNTRY : China
GROUP : Product Group

我们希望 IMOS 不仅是一款易用、美观的车机系统，更希望它为用户带来更丰富、更有温情的人文关怀和情感体验。

We hope the IMOS is not only an easy-to-use and beautiful Information System HMI, also a richer and more warm humanistic care and emotional experience for users.

科大讯飞智慧黑板
IFLYTEK SMART BLACKBOARD

作者: 雷辰阳 诸臣 许宝月
机构: 科大讯飞股份有限公司
国家: 中国
组别: 产业组

AUTHOR : Lei Chenyang, Zhu Chen, Xu Baoyue
UNIT : IFLYTEK Co., Ltd.
COUNTRY : China
GROUP : Product Group

这是一款通过音视频技术、语音识别等方式解决老师在教学环节中负担重、效率低问题的课堂教学产品。

This is a smart teaching product that solves the problem of heavy burden and low efficiency for teachers in the teaching process through audio and video technologies, voice recognition and other methods.

智能编程机器车
U-CAR

作者: 赵鸿 诸臣 许宝月
机构: 科大讯飞股份有限公司
国家: 中国
组别: 产业组

AUTHOR : Zhao Hong, Zhu Chen, Xu Baoyue
UNIT : IFLYTEK Co., Ltd.
COUNTRY : China
GROUP : Product Group

这是一款集成多种传感器的智能驾驶教学套件,学生可轻松获取点云和图像数据,进行编程和算法的实践学习。

This is a smart driving teaching kit integrating various sensors. Students can easily obtain point cloud and image data for the application and algorithm practice of programming practice learning.

网易数字文化中心
NETEASE DIGITAL CULTURE CENTER

作者: 林智 袁思思 顾费勇 刘勇成 胡志鹏
机构: 网易雷火 UX
国家: 中国
组别: 产业组

AUTHOR : Lin Zhi, Yuan Sisi, Gu Feiyong, Liu Yongcheng, Hu Zhipeng
UNIT : NetEase ThunderFire UX
COUNTRY : China
GROUP : Product Group

一款以中国传统文化素材库为功能核心的网易数字文化中心 PC 官网。

A PC official website of NetEase Digital Cultural Center with Chinese traditional cultural material library as the core function.

BEYOND

作者: Zhuoneng Wang, Greg Chen, Youngryun Cho, Wei-Chieh Wang
国家: 美国
组别: 概念组

AUTHOR : Zhuoneng Wang, Greg Chen, Youngryun Cho, Wei-Chieh Wang
COUNTRY : United States
GROUP : Concept Group

BEYOND 是一种未来的混合现实博物馆体验，以声音、眼神和手势等多感官互动为基础，为游览者提供身临其境、引人入胜的旅程。

BEYOND is a future mixed-reality museum experience offering an immersive and engaging journey rooted in multi-sensory interactions, such as voices, eyes, and gestures.

(ID)ENTITIES

SOUNDS OF FREEDOM

作者：Chukwuma Anagbado, Antonia Kihara
国家：尼日利亚
组别：产业组

AUTHOR : Chukwuma Anagbado, Antonia Kihara
COUNTRY : Nigeria
GROUP : Product Group

作者：Mutana Wanjira Gakuru, Victor Ndisya
机构：Fiction Entertainment / African Fiction Academy
国家：肯尼亚
组别：产业组

AUTHOR : Mutana Wanjira Gakuru, Victor Ndisya
UNIT : Fiction Entertainment / African Fiction Academy
COUNTRY : Kenya
GROUP : Product Group

Lizaad 是一个提供家具和生活方式产品的图案艺术品牌。我们推广来自非洲的具有丰富文化内涵的设计实践。该品牌的定位是构思、设计、策划和零售以非洲为灵感的产品——美学、传统、功能、材料、民间传说、文化和非洲特色。

Lizaad is a pattern art brand offering furniture and lifestyle products. We promote the culturally rich design practice from Africa. The brand is positioned to conceive, design, curate and retail products that are inspired by Africa – aesthetics, heritage, function, materials, folklore, culture and African identity.

Sounds of Freedom 是一款产品，为数百万寻求通过沉浸式体验，重新发现其文化遗产的非洲人打开了接触文化之门。只需下载应用程序，选择您的探险体验，然后按下播放键即可。

Sounds of Freedom is a product that opens access to culture for millions of Africans seeking to rediscover their cultural heritage through immersive experiences. Just download the app, choose your adventure, and press play.

联想备授课
LENOVO TEACHING ALL-IN-ONE

作者：沈文京 尹婕 姚涔 安尉 蒙骁
机构：联想研究院
国家：中国
组别：产业组

AUTHOR : Shen Wenjing, Yin Jie, Yao Cen, An Wei, Meng Xiao
UNIT : Lenovo Research
COUNTRY : China
GROUP : Product Group

联想备授课是一款专业服务教师的软硬一体化方案，凭借其出众的硬件和丰富的软件生态，帮助教师减轻备课负担，提升教学效率。

Lenovo Teaching All-in-One is a total solution with software and hardware integration for professional serving educators. With its outstanding hardware and rich software ecology, Teaching All-in-One helps teachers reduce the burden of lesson preparation and improve teaching efficiency.

妙笔
MAGIC BRUSH

作者：陈舒窈 张颖 徐浩然
机构：浙江大学
国家：中国
组别：概念组

AUTHOR : Chen Shuyao, Zhang Ying, Xu Haoran
UNIT : Zhejiang University
COUNTRY : China
GROUP : Concept Group

妙笔是一款专注于新国风绘画创作、通过 AI 图像生成技术降低新手作画门槛的辅助绘画工具。

Magic Brush is an auxiliary painting tool that focuses on new national style painting creation. It helps to reduce the threshold of novice painting by AI image generation technology.

大数据领域全链路数据治理设计重塑
DESIGN RESHAPING OF END-TO-END BIG DATA GOVERNANCE

TWINZO

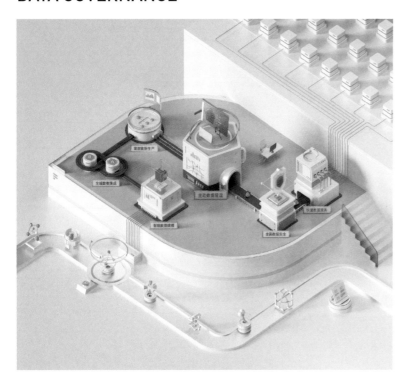

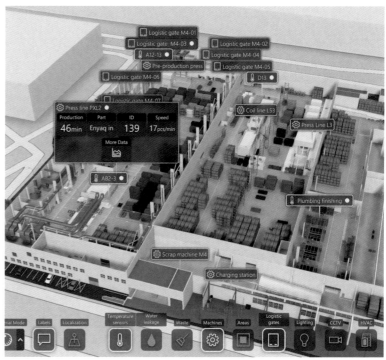

作者: 陈磊 朱凤瑜 蒋贝妮 胡永棋方
机构: 阿里云计算有限公司
国家: 中国
组别: 产业组

AUTHOR: Chen Lei, Zhu Fengyu, Jiang Beini, Hu Yongqifang
UNIT: Alibaba Cloud Computing Co.,Ltd.
COUNTRY: China
GROUP: Product Group

作者: Michal Ukropec, Jiri Zila, Michal Celeng, Patrik Pasko, Tomas Vojtek
机构: 5.0 technologies j.s.a.
国家: 斯洛伐克共和国
组别: 产业组

AUTHOR: Mutana Wanjira Gakuru, Victor Ndisya
UNIT: 5.0 technologies j.s.a.
COUNTRY: The Slovak Republic
GROUP: Product Group

DataWorks 是阿里巴巴十二年数据开发与治理的最佳实践，在数字化转型场景内，作为大数据中台最重要的平台型产品。

DataWorks becomes the best practice in the development and management of Alibaba in the past 12 years.It is the most significant platform-based product for big data mid-end within the digital transformation scenario.

twinzo 是一款3D 实时数字孪生产品，为全球工厂的移动和桌面用户带来了独特的可视化和集成工具。首先，上传模型、整合数据、下载应用程序并查看工厂的情况 —— 物流、生产、质量或环境数据。之后，进行预测、模拟、比较。

twinzo, the 3D live digital twin, brings a unique visualization and integration tool for mobile and desktop users for factories worldwide. First, upload the model, integrate data, download the app and see what's going on at your plant - logistics, production, quality, or environmental data. Later, predict, simulate, compare.

百度百变人生
BAIDU'S MINI PROGRAM -COLORFUL LIVES

作者: 史玉洁 张勇 刘思任 李爽 陈映钐
机构: 百度在线网络技术 (北京) 有限公司
国家: 中国
组别: 产业组

AUTHOR : Shi Yujie, Zhang Yong, Liu Siren, Li Shuang, Chen Yingshan
UNIT : Baidu Online Network Technology (Beijing) Co., Ltd.
COUNTRY : China
GROUP : Product Group

这是一款基于百度 APP 的智能职业体验平台,通过趣味性的体验形式,帮助用户探索更多的职业可能性。

This is a smart vocational experience platform based on Baidu APP, which draws on a form of interesting experience to help users explore more career possibilities.

小爱同学个人定制智能助手
XIAOAITONGXUE PERSONAL AI INTELLIGENT ASSISTANT

作者: 刘静 贾雪威 伍仪华 薛骁 林兆梅
机构: 北京小米松果电子有限公司
国家: 中国
组别: 产业组

AUTHOR : Liu Jing, Jia Xuewei, Wu Yihua, Xue Xiao, Lin Zhaomei
UNIT : Beijing Pinecone Electronics Co., Ltd.
COUNTRY : China
GROUP : Product Group

小爱同学是内置于小米的手机及智能设备中的,基于语音交互,结合 AI 驱动的自定义虚拟形象的智能助手软件服务。

Xiaoaitongxue is a personal AI assistant built in Xiaomi smartphone, television, smart speaker and other devices. We employ our technology of virtual avatar engine to give a personified digital image to the Voice Interaction with which users can interact naturally and intuitively.

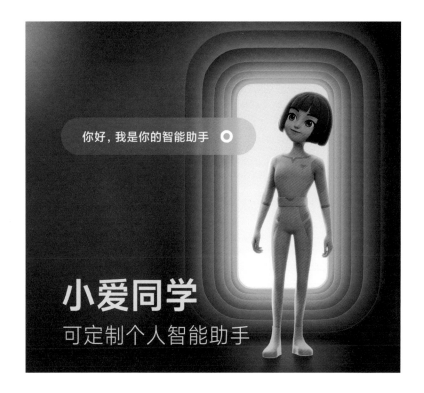

网易瑶台 —— 会展产业数字化跃迁
NETEASE YAOTAI — ADVANCED DIGITAL CONFERENCE & EXPO

作者：郭冠敏 刘昊 曾靖喜 曹力文 陈雅萍
机构：网易伏羲
国家：中国
组别：产业组

AUTHOR : Guo Guanmin, Liu Hao, Zeng Jingxi, Cao Liwen, Chen Yaping
UNIT : FuXi
COUNTRY : China
GROUP : Product Group

网易瑶台，提供基于 AI 的"场景、虚拟人、富媒体交互"的数字空间，实现会展产业后疫情时代的数字化跃迁。

Based on AI technology, Netease Yaotai provides immersive digital event spaces for enterprises and users through digital environments, virtual characters and rich media social interactions, to realize the digital transition of conference and exhibition industry (C&E industry) in the post-COVID-19 era.

京东智造云 AI 决策仿真平台
AI-BASED DECISION — MAKING SIMULATION PLATFORM

作者：周澍 迟李青 吕昊 曹铭喆 胡炜
机构：京东科技信息技术有限公司
国家：中国
组别：产业组

AUTHOR : Zhou Shu, Chi Liqing, Lv Hao, Cao Mingzhe, Hu Wei
UNIT : JD Technology Information Technology Co., Ltd.
COUNTRY : China
GROUP : Product Group

京东智造云通过 AI 决策仿真平台，让柔性制造成为可能，在不确定性中发现并主动寻找最优计划。

AI-based Decision-making Simulation Platform, powered by JD Cloud, is committed to build a whole-chain digital twin of flexible manufacturing, on which enterprises can synchronize factories and markets, seek and implement the best plan under uncertainties.

格力云数字孪生平台
GREE CLOUD DIGITAL TWIN FACTORY

作者：赖元杰 熊文彬 张法祥 郭正圻 梁根蔚
机构：格力电器股份有限公司
国家：中国
组别：产业组

AUTHOR : Lai Yuanjie, Xiong Wenbin, Zhang Faxiang, Guo Zhengqi, Liang Genwei
UNIT : Gree Electric Appliances,Inc.
COUNTRY : China
GROUP : Product Group

通过构建智能智制工厂场景3D可视化解决方案，实现数字化工厂生产的可视化管理，全面提升决策管理效率。

By building a 3D visualization platform solution for intelligent factory scenes, we can realize the visualization management of digital factory production and comprehensively improve the efficiency of decision making.

AKILA DIGITAL TWIN PLATFORM

作者：Wilfred Leung Wei Lit
机构：Akila
国家：新加坡
组别：产业组

AUTHOR : Wilfred Leung Wei Lit
UNIT : Akila
COUNTRY : Singapore
GROUP : Product Group

一个数字孪生平台，用于优化建筑物的环境影响和运营，并简化 ESG 报告。

A digital twin platform to optimize the environmental impact and operation of buildings and streamline ESG reporting.

讯飞智能办公本 UI 交互系统
IFLYTEK AINOTE UI INTERACTIVE SYSTEM

作者：张晗 李守强 程琛 吴蒙勤 肖栋添
机构：合肥讯飞读写科技有限公司
国家：中国
组别：产业组

AUTHOR : Zhang Han, Li Shouqiang, Cheng Chen, Wu Mengqin, Xiao Dongtian
UNIT : iFLYINK
COUNTRY : China
GROUP : Product Group

科大讯飞人工智能语音技术赋能办公的交互系统，让办公人群更专注于信息理解与深度思考，提升办公效率。

IFLYTEK's interactive office system enabled by artificial intelligent voice technology enables office workers to focus more on information understanding and in-depth thinking to improve office efficiency.

SMART CHILDCARE CENTER / LOOKMEE

Facial Recognition

Object and eye Recognition

Scene Analysis

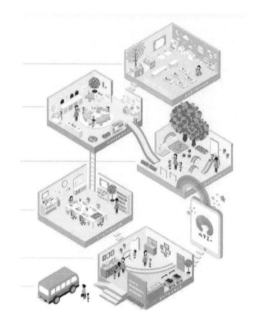

作者：Yasuyuki Toki, Akiko Asano, Hiroaki Akanuma
机构：Unifa Inc
国家：日本
组别：产业组

AUTHOR : Yasuyuki Toki, Akiko Asano, Hiroaki Akanuma
UNIT : Unifa Inc
COUNTRY : Japan
GROUP : Product Group

"智慧托育中心"是托育中心的未来愿景，其结合我们利用人工智能和物联网等尖端技术开发的软件包 "Lookmee"，旨在减轻托育工作人员的工作量，提高托育质量，并最终提升家庭幸福。

"Smart Childcare Center" is a future vision of childcare centers incorporating "Lookmee", a software package which we have developed with the cutting-edge technologies, such as AI and IoT, to reduce the workload of childcare workers, improve the quality of childcare, and eventually encourage happiness for families.

A-VIBE

作者：Tianqin Lu
国家：荷兰
组别：概念组

AUTHOR : Tianqin Lu
COUNTRY : Dutch
GROUP : Concept Group

A-Vibe 是一种非实时的动物造型虚拟形象系统，旨在将用户当前的真实状态转化为定制的动物造型，从而有助于在线学习环境中保持社交存在感和连通性。（此内容使用 Live2D Inc. 公司拥有并受版权保护的示例数据。）

A-Vibe as a non-real-time animal-form avatar system has been created to work to translate the user's current honest state into a customised animal form, thus contributing to social presence and connectedness in the online learning environment. (This content uses sample data owned and copyrighted by Live2D Inc.)

与菌绝 - 针对酒店隔离人员的服务设计
SERVICE DESIGN FOR HOTEL ISOLATION PERSONNEL

作者：陈天易 胡超杰 张琪 刘豪 李熠炫
机构：浙江工业大学
国家：中国
组别：概念组

AUTHOR : Chen Tianyi, Hu Chaojie, Zhang Qi, Liu Hao, Li Yixuan
UNIT : Zhejiang University of Technology
COUNTRY : China
GROUP : Concept Group

与菌绝是一套通过智能物联网，改善隔离人员酒店隔离体验的服务设计。

YuJunjue is a set of services designed to improve the isolation experience in hotels through the intelligent Internet of Things.

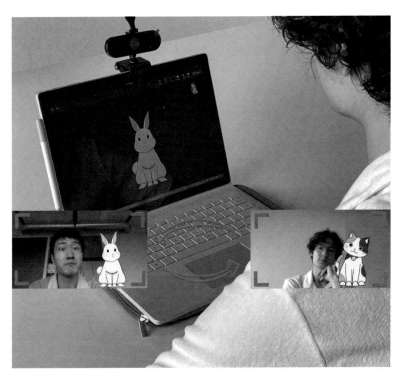

COCOTRUCK

作者：Marina Kim, Seunghoon Jeong
机构：COCONUT SILO
国家：韩国
组别：产业组

AUTHOR : Marina Kim, Seunghoon Jeong
UNIT : COCONUT SILO
COUNTRY : South Korea
GROUP : Product Group

COCOTRUCK 是一个人工智能数字物流平台，聚集了物流市场的各类参与者，使用户能够最大限度地提高效率，同时减少温室气体排放。

COCOTRUCK is an AI digital logistics platform that gathers all types of players in the logistics market, enabling users to maximize efficiency and reduce greenhouse gas emission.

医疗第三空间
MEDICAL THIRD SPACE

作者：张明鈜 陈玛丽
国家：英国
组别：概念组

AUTHOR : Zhang Minghong, Chen Mali
COUNTRY : UK
GROUP : Concept Group

医疗第三空间通过"虚拟"隐私空间提供整合医疗服务。

Medical Third Space is a "virtual" privacy space to provide integrated medical services.

斑马数智化导览服务
BANMA INTELLIGENT DIGITAL GUIDER SERVICE

作者：李经 胡铭洋 余晓瑜 陈思承 张立
机构：斑马网络技术有限公司
国家：中国
组别：产业组

AUTHOR : Li Jing, Hu Mingyang, Yu Xiaoyu, Chen Sicheng, Zhang Li
UNIT : Banma Network Technology Co.,Ltd.
COUNTRY : China
GROUP : Product Group

斑马数智化导览服务是依托大数据平台，结合自动驾驶技术，为游客出游提供无接触式导览服务。

The contactless intelligent guiding service provided by Banma digital intelligent guiding service relies on a big data platform and combines self-driving technique.

SEEPORT 室内交互型 VR 运动设备
SEEPORT — INDOOR INTERACTIVE VR SPORTS EQUIPMENT

作者：李玉媚 林安琪 江靓
国家：中国
组别：概念组

AUTHOR : Li Yumei, Lin Anqi, Jiang Liang
COUNTRY : China
GROUP : Concept Group

Seeport 是一款通过 VR 设备解决相隔较远的人们无法一起进行各类运动的一款家庭数字健身娱乐产品。

Seeport is a family digital fitness and entertainment product that solves the problem that people who are far apart cannot play all kinds of sports together through VR devices.

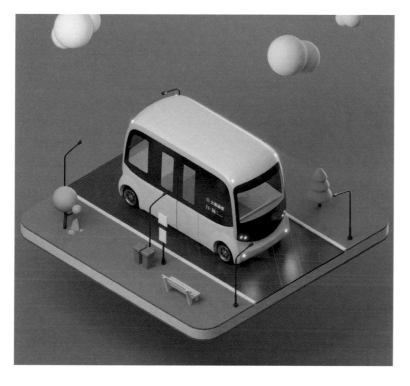

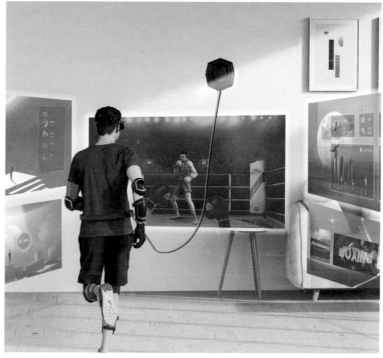

百度知识胶囊
KNOWLEDGE CAPSULES

CULTURAL ADVOCATE PROGRAM

作者：史玉洁 张勇 韩璐 李浩 欧阳石
机构：百度在线网络技术（北京）有限公司
国家：中国
组别：产业组

AUTHOR : Shi Yujie, Zhang Yong, Han Lu, Li Hao, Ouyang Shi
UNIT : Baidu Online Network Technology (Beijing) Co., Ltd.
COUNTRY : China
GROUP : Product Group

作者：José Pablo Domínguez, Tako Sulakvelidze, Claire Yuan Zhuang, Valentina Palacios, N/A
国家：美国
组别：概念组

AUTHOR : José Pablo Domínguez, Tako Sulakvelidze, Claire Yuan Zhuang, Valentina Palacios, N/A
COUNTRY : United States
GROUP : Concept Group

通过百度专业化且丰富的内容库，知识胶囊心理健康小程序通过 AI 助手帮助患有心理疾病的用户调节心理健康。

Based on the specialized and rich database of Baidu, Knowledge Capsules - Mental Health Smart Program help to enhance mental health of users with mental diseases via the AI assistant.

为了减少美国医疗系统中的沟通障碍、语言障碍和文化误解，拥有不同文化背景的患者可以与来自相似地区的倡导者进行匹配，以了解他们的需求并提供足够的帮助。

To reduce communication gaps, language barriers, and cultural misunderstandings in the USA healthcare system, patients from diverse cultural backgrounds can be matched with advocates from similar regions to understand their needs and provide adequate assistance.

基于视觉 AI 算法的医疗行为智能分析与数字化管理系统
MEDICAL BEHAVIOR ANALYSIS MANAGEMENT SYSTEM

商汤绝影智能车舱
SENSEAUTO CABIN

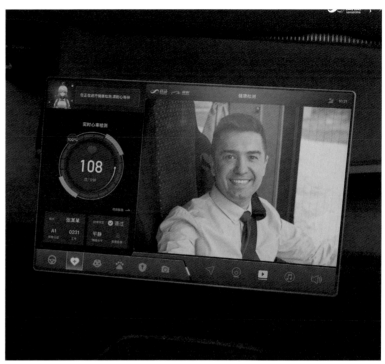

作者：应东东 赵凯 屈世豪 范育芳 董少谦
机构：杭州百世伽信息科技有限公司 / 杭州哲美文化产业发展有限公司
国家：中国
组别：产业组

作者：唐御聪 孙康 闫雪 李轲
机构：北京商汤科技开发有限公司
国家：中国
组别：产业组

AUTHOR : Ying Dongdong, Zhao Kai, Qu Shihao, Fan Yufang, Dong Shaoqian
UNIT : Hangzhou Bestplus Information Technology Co., Ltd. / Hangzhou Zhemei Cultural Industry Development Co., Ltd.
COUNTRY : China
GROUP : Product Group

AUTHOR : Tang YuCong, Sun Kang, Yan Xue, Li Ke
UNIT : Sensetime
COUNTRY : China
GROUP : Product Group

对医疗行为进行实时监测、预警、分析，为医疗机构医疗行为管理与质量改进提供数字化决策依据。

Real-time monitoring, early warning, and analysis of medical behaviors as well as provide digital decision-making basis for medical behavior management and quality improvement in medical institutions.

本产品为车辆驾驶者及乘员打造，采用 AI 技术监测疲劳、注意力、健康状态等，为交通安全和身体健康保驾护航。

This product is designed for vehicle drivers and occupants, and uses AI technology to monitor fatigue, attention, health status, etc., to protect traffic safety and physical health.

AED 除颤仪的 AR 培训系统设计
AR TRAINING SYSTEM DESIGN OF AED DEFIBRILLATOR

SKOP

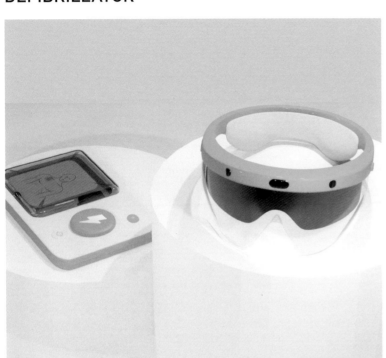

作者：王小东
机构：浙江大学
国家：中国
组别：概念组

AUTHOR : Wang Xiaodong
UNIT : Zhejiang University
COUNTRY : China
GROUP : Concept Group

作者：LECROQ Cyrille, ARNAUD Philippe, BERNELIN Alexandre
机构：WEMED Studio PAD
国家：法国
组别：产业组

AUTHOR : LECROQ Cyrille, ARNAUD Philippe, BERNELIN Alexandre
UNIT : WEMED Studio PAD
COUNTRY : France
GROUP : Product Group

AED 除颤仪的 AR 培训系统是一款通过 AR 技术来解决疫情常态化下高校无法大规模线下培训的痛点，采用预约制 AR 培训。

The AR training system of AED defibrillator is a way to solve the pain points that colleges and universities cannot conduct large-scale offline training under the normalization of the epidemic through AR technology. It adopts the appointment system AR training.

SKOP 是一款3D 打印的联网听诊器，可以让那些不能离开家或生活在电网之外的人享受身临其境的医生听诊。

SKOP is a 3D printed connected stethoscope that allows people who can't leave home or live off the grid to get a real auscultation from a doctor.

阿里云数字医疗智能应用
ALIBABA CLOUD DIGITAL HEALTHCARE INTELLIGENT SYSTEM

作者：廖敏月 许赞 刘漪 张惠顺 叶旭辉
机构：阿里云计算有限公司
国家：中国
组别：产业组

AUTHOR : Liao Minyue, Xu Zan, Liu Yi, Zhang Huishun, Ye Xuhui
UNIT : Alibaba Cloud Computing Co.,Ltd.
COUNTRY : China
GROUP : Product Group

阿里云数字医疗智能应用，优化资源配置、提供精准决策，为医疗生态保驾护航，让市民享受最佳医疗服务。

Alibaba Cloud digital medical intelligent application aims to provide the best medical services for citizens through optimizing resource allocation, providing accurate decision-making, escorting the medical ecology.

华为生态通
ECOCAPTAIN

作者：杨敏敏 唐小敏 陈海武
机构：华为技术有限公司
国家：中国
组别：产业组

AUTHOR : Yang Minmin, Tang Xiaomin, Chen Haiwu
UNIT : Huawei Technologies Co.,Ltd.
COUNTRY : China
GROUP : Product Group

华为生态通是数字化生态治理综合解决方案，系统及时识别企业污染物排放异常，自动上报相应的处置部门。

A mobile workbench station based on cloud service that helps civil servant manage and handle the environmental emergency.

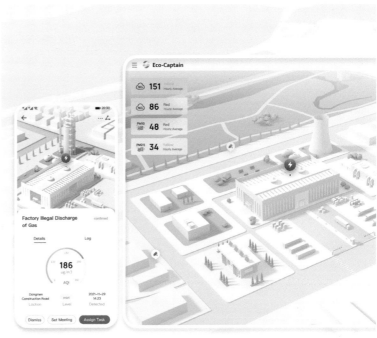

NUROX

作者：EOJIN JUN
国家：韩国
组别：概念组

AUTHOR：EOJIN JUN
COUNTRY：South Korea
GROUP：Concept Group

"NUROX"是一种全面的物流流动性系统，解决了包裹递送浪费问题，并保障了递送司机的人权，同时实现可持续流动。

"NUROX"is a holistic logistics mobility that solves the problem of parcel delivery waste and the human rights of delivery drivers for sustainable flow.

商用清洁机器人及软件平台
COMMERCIAL CLEANING ROBOT AND SOFTWARE PLATFORM

作者：陆益彬 宋志敏
机构：美的集团 - 美智纵横科技有限公司
国家：中国
组别：产业组

AUTHOR：Lu Yibin, Song Zhimin
UNIT：Midea Group
COUNTRY：China
GROUP：Product Group

商用清洁机器人及软件平台，集成商用场景的服务机器人，通过智能化的方式缓解商用清洁市场的人力不足的矛盾。

Commercial cleaning robots and platforms, integrated service robots for commercial scenarios, alleviate the contradiction of insufficient manpower in the commercial cleaning the market after the aging population through intelligent methods, and provide a more free and convenient future lifestyle through technology.

微循环快充充电架
MICRO-CIRCULATION FAST CHARGING CHARGER

作者：乔楠 刘威 金德智 范志航 程胜超
机构：宇通客车股份有限公司 / 郑州飞鱼设计有限公司
国家：中国
组别：产业组

AUTHOR : Qiao Nan, Liu Wei, Jin Dezhi, Fan Zhihang, Cheng Shengchao
UNIT : Yutong Bus Co.,Ltd./ Feish Design（Zhengzhou）Co.,Ltd.
COUNTRY : China
GROUP : Product Group

微循环快充充电架是一款利用高压快充技术为无人驾驶微循环车辆提供快速充电服务的公共设施。

The micro-cycle fast-charging rack is a public facility that uses high-voltage fast-charging technology to provide fast-charging services for unmanned micro-cycle vehicles.

斐视 5G 远程驾驶舱
FISON 5G COCKPIT

作者：周才致 田野 张沙 胡圣贤 何遥
机构：长沙斐视科技有限公司
国家：中国
组别：产业组

AUTHOR : Zhou Caizhi, Tian Ye, Zhang Sha, Hu Shengxian, He Yao
UNIT : FISON Technology Co., Ltd.
COUNTRY : China
GROUP : Product Group

斐视驾驶舱是一款通过 5G 远程操控技术，解决工程机械在执行高危任务时人员易发生安全事故问题的智能装备。

The FISON Cockpit is an intelligent device that uses 5G remote control technology to solve the problem that construction machinery is prone to safety accidents when performing high-risk tasks.

HERDVISION

作者 : Matthew Dobbs, Ben McCarthy, Simon McDonald, Heather Sanders
机构 : Agsenze Ltd.
国家 : 英国
组别 : 产业组

AUTHOR : Matthew Dobbs, Ben McCarthy, Simon McDonald, Heather Sanders
UNIT : Agsenze Ltd.
COUNTRY : UK
GROUP : Product Group

HerdVision 是一种创新的农业技术系统，通过每日自动监测奶牛的健康状况来提高农场效率和奶牛健康。该技术使用 2D/3D 摄像头、机器学习、本地和基于云的算法的独特组合，通过应用程序、网络或现有农场管理软件生成自动化健康报告。

HerdVision is an innovative AgTech system that improves farm efficiency and cow welfare through automatic daily monitoring of cow health. Using a unique combination of 2D/3D cameras, machine learning, local and cloud based algorithms, the technology produces automated health reports through app, web or existing farm management software.

COFFEED

作者 : Zongheng Sun, Yumeng Li
机构 : LI & SUN DESIGN LLC/PEAR & MULBERRY
国家 : 美国
组别 : 概念组

AUTHOR : Zongheng Sun, Yumeng Li
UNIT : LI & SUN DESIGN LLC/PEAR & MULBERRY
COUNTRY : United States
GROUP : Concept Group

CofFeed 是一个协调循环农业有机土壤收集的平台。经过后处理，废咖啡渣（SCG）变成了一种更好的肥料，从而消除了肥力损失、空气污染甚至健康风险。该平台通过集体努力将分散的城市食物浪费转化为郊区的有机肥料。

CofFeed is a platform that coordinates organic grounds collection for circular farming. After post-processing, spent coffee grounds (SCG) turn into a better fertilizer that eliminates fertility loss, air pollution, and even health risks. It transforms scattered urban food-wasting into suburban organic fertilizing by engaging collective efforts.

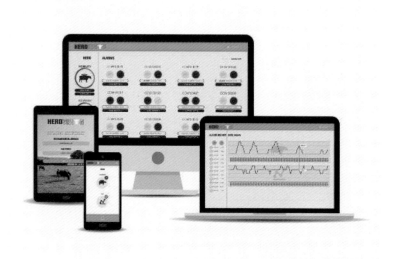

美啊设计平台 —— 培养引领未来设计师的在线教育平台
MEIA DESIGN PLATFORM— AN ONLINE EDUCATION PLATFORM

VUVU 生物教具
VUVU

作者：胡晓 苏菁 胡蓉 罗志国
机构：广州美啊教育有限公司
国家：中国
组别：产业组

AUTHOR : Hu Xiao, Su Jing, Hu Rong, Luo Zhiguo
UNIT : Guangzhou Meia Education Co. , Ltd.
COUNTRY : China
GROUP : Product Group

作者：杨俊辉 卓宜萱 简邱伟 许品宜 陈沛颐
机构：NetEase ThunderFire UX
国家：中国
组别：产业组

AUTHOR : Yang Junhui, Zhuo Yixuan, Jian Qiuwei, Xu Pinyi, Chen Peiyi
UNIT : NetEase ThunderFire UX
COUNTRY : China
GROUP : Product Group

美啊设计平台是一款通过提供设计前瞻知识与案例，解决设计师专业困惑与瓶颈的数字化在线教育服务平台。

MEIA design platform is a digital online education service platform that solves the professional confusion and bottleneck of designers by providing design-forward knowledge and cases.

一款探究式生物数位游戏教具，从"学习既有知识"为主的学习，转为强调实作寻求解答"创造知识"的体验。

A digital teaching aid for learning biological. By simulating the circumstances in the game, explorative learning is expected to be effectively integrated into scientific courses, transforming the main learning method of "learning the existing knowledge" into the immersive learning of "creating knowledge" that emphasizes on practice and solution finding.

ZHY963 ——智慧融合可视化终端
ZHY963 — INTELLIGENT FUSION VISUAL TERMINAL

第四范式 AIOT —— 智慧园区综合决策系统
AIOT — SMART PARK INTEGRATED DECISION-MAKING SYSTEM

作者：赵砚青 刘庆超 范仕亮 王明立 杨帆
机构：智洋创新科技股份有限公司
国家：中国
组别：产业组

AUTHOR : Zhao Yanqing, Liu Qingchao, Fan Shiliang, Wang Mingli, Yang Fan
UNIT : Zhiyang Innovation Technology Co., Ltd.
COUNTRY : China
GROUP : Product Group

作者：迟娩 王凯 徐昀 王蔚
机构：第四范式（北京）技术有限公司
国家：中国
组别：产业组

AUTHOR : Chi Mian, Wang Kai, Xu Yun, Wang Wei
UNIT : Fourth Paradigm (Beijing) Data & Technology Co., Ltd.
COUNTRY : China
GROUP : Product Group

ZHY963是一款架设在输电线路杆塔上的智能监拍装置，通过设备前端 AI 计算，解决输电线路的隐患识别、告警等问题。

ZHY963 is an intelligent monitoring device erected on the transmission line tower. Through the front-end AI calculation, the equipment solves the hidden danger identification, alarm and other problems of the transmission line.

第四范式智慧园区综合决策系统 +AIOT 平台，利用 AI 技术推动新一代科技、高效、安全的智慧园区发展，普惠各行各业。

The Fourth Paradigm Smart Park integrated decision-making system +AIOT platform uses AI technology to promote the development of a new generation of scientific, efficient, and safe Smart Parks, benefiting all walks of life.

CIRCLE SAVINGS

作者：Patrick Awori
机构：Imaginarium
国家：肯尼亚
组别：产业组

AUTHOR：Patrick Awori
UNIT：Imaginarium
COUNTRY：Kenya
GROUP：Product Group

Circle 让支付和省钱变得简单、实惠和透明。任何团体或个人都可以在几秒钟内从世界任何地方随时支付或储蓄任何金额。我们会在你每次储蓄时奖励奖金，以帮助您实现财务目标。

Circle makes payments and saving money easy, affordable and transparent. Any group or individual can pay or save any amount, at any time, from anywhere in the world in a matter of seconds. We reward bonus money each time you save, so you can reach your financial goals.

城市与我 —— 创新型城市数字孪生系统
CITY&ME — INNOVATIVE DIGITAL TWIN SYSTEM OF CITY

作者：詹明明 王思越 苗雨菲 陈俊锰 黎咸训
机构：中国电子系统技术有限公司
国家：中国
组别：产业组

AUTHOR：Zhan Mingming, Wang Siyue, Miao Yufei, Chen Junmeng, Li Xianxun
UNIT：China Electronic System Technology Co., Ltd.
COUNTRY：China
GROUP：Product Group

城市数字孪生系统是一款通过对城市实时数据进行分析映射，以辅助城市指挥调度或决策的计算机大屏展示系统。

Urban digital twin system is a computer large screen display system that can assist urban command and dispatching decision-making through real-time display and analysis of urban data information.

腾讯技术公益志愿者平台
T-VOLUNTEER NETWORK

作者：邱颖彤 郑露 张翘 陈晓畅 饶瑞
机构：腾讯科技（深圳）有限公司
国家：中国
组别：产业组

AUTHOR : Qiu Yingtong, Zheng Lu, Zhang Qiao, Chen Xiaochang, Rao Rui
UNIT : Tencent Technology (Shenzhen) Co.,Ltd.
COUNTRY : China
GROUP : Product Group

通过建设面向公益行业的数字志愿服务平台，提升公益工作效率，旨在为公益机构和专业志愿者搭起沟通桥梁。

In order to improve the efficiency of public welfare work, the project builds a digital volunteer service platform for the public welfare industry, which also provides a bridge of communication between public welfare organizations and professional volunteers.

智能矿山
INTELLIGENT MINING

作者：高健 杨敏敏 文梦芝 柳玮 朱磊
机构：华为技术有限公司
国家：中国
组别：产业组

AUTHOR : Gao Jian, Yang Minmin, Wen Mengzhi, Liu Wei, Zhu Lei
UNIT : Huawei Technologies Co.,Ltd.
COUNTRY : China
GROUP : Product Group

华为智能矿山通过5G、云、AI和大数据等技术，解决安全、增效等问题，构建万物互联的智能矿山。

By using 5G, cloud, AI, and big data technologies, Huawei intelligent mining solutions help customers build intelligent mines where all things are connected and operations are more efficient and safer.

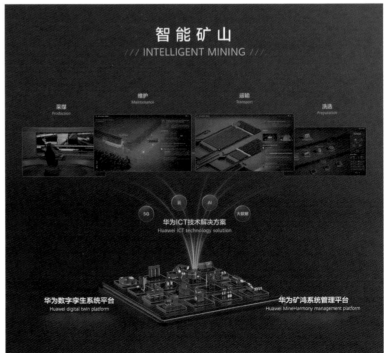

BOTTO! ——THE IOT DEVICE AND CHATBOT AGAINST FOOD WASTE OPENDOT

PLANTIVERSE

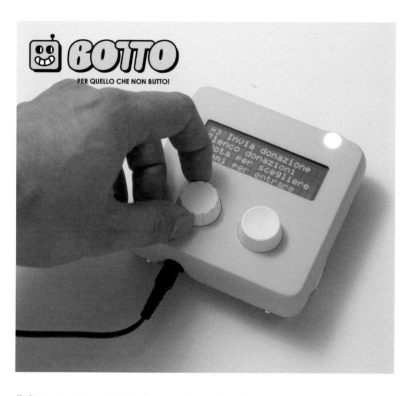

作者 : Enrico Bassi, Vittorio Cuculo, Antonio Garosi
机构 : OpenDot
国家 : 意大利
组别 : 产业组

AUTHOR : Enrico Bassi, Vittorio Cuculo, Antonio Garosi
UNIT : OpenDot
COUNTRY : Italy
GROUP : Product Group

作者 : Cecilia Tham, Mark Bunger, Magda Mojsiejuk, Nuria Albo,David Tena
机构 : Futurity Systems
国家 : 西班牙
组别 : 概念组

AUTHOR : Cecilia Tham, Mark Bunger, Magda Mojsiejuk, Nuria Albo,David Tena
UNIT : Futurity Systems
COUNTRY : Spain
GROUP : Concept Group

BOTTO 是一种物联网技术设备, 它可以与聊天机器人进行通信, 从而促进批发商和捐赠剩余食物的协会之间的沟通。它的设计和开发是为了使城市更具可持续性, 可用于报告和干预城市和城市周边地区的食物流动。

BOTTO is an IoT technology device which communicates with a chatbot that facilitates the communication among wholesalers and associations that act for donation of surplus food. It has been designed & developed to make the city more sustainable, reporting and intervening on the food flow in urban & peri-urban areas.

Plantiverse 是我们尝试通过创建真实植物、树木和其他生态系统的数字孪生体, 来创建一个物种间的元宇宙, 并将其转换为 NFTrees, 每个 NFTrees 都由植物设计, 为植物造福, 并在碳中和区块链中播种, 为可持续计划提供资金。

Plantiverse is our attempt to create an interspecies metaverse to live by creating digital twins of real plants, trees and other ecology, converted into NFTrees, each designed by plants, for the benefit of plants, seeded in a carbon neutral blockchain to fund sustainable initiatives.

关于 DIA
ABOUT DIA

大奖缘起

2014年10月"最设计"展览期间李强同志与中国美术学院许江老院长提出创办国际级工业设计大奖的构想，并决定由中国美术学院主办。2016年1月8日，袁家军同志宣布中国设计智造大奖 (Design Inteligence Award, DIA) 正式启动。

2014年10月"最设计"展览现场
"Design Design" Exhibition in October 2014

2016年1月8日，袁家军同志宣布首届 DIA 启动并面向全球发布"作品征集令"。
On January 8, 2016, Comrade Yuan Jiajun announced the official launch of DIA and extended a global call for entries.

大奖定位

中国设计智造大奖是2015年创立的当代创新设计评价、推广与合作平台，是一个重塑当代设计标准的新型国际工业设计奖项， 也是一个艺术、科技与商业的跨界创新全球竞赛， 更是一个创意转向财富与未来的实体创新加速器。

大奖价值观

以"人文智性、生活智慧、科艺智能、产业智库"为核心价值观，倡导设计回归"智造"本源。这是一个由中国发起、全球共享的设计大赛，设计的本质就是为人、为生活，"茶米为食、麻丝为衣、竹陶为用、林泉为居"映射出东方物质文明的可持续发展理念，蕴藏着设计的生活智慧与造物方式，"智"乃智巧、智能、智慧、智性 …… 知晓天地人事规律与创造方法相应之义，"造"乃造型、造物、造境、造化 …… 映射出中国悠久的文化与大设计视野，故，设计是"智"与"造"的融会贯通，即"智造"。我们以中国智慧，重新审视源于西方工业文明的现代设计概念，倡导"智造"引领新时代的设计行业发展。

评审机制

设初评、复评、总决赛三轮评审环节。初评采用图文评审方式，复评采用实物结合视频的评审方式，总决赛采用参赛选手现场答辩的评审方式。大奖集聚设计大师与跨界专家构成国际一流专家库，实行提名专家与评审专家互相独立的工作机制。

评审标准

大奖立足智能制造大时代背景，独创"金智塔"评价体系，包含三层标准：
一是基础标准，强调"设计之技"，包含功能性、美学性、技术性、体验性、可持续性等评价因子；
二是核心标准，强调"设计之道"，包含民生贡献度、产业贡献度和未来贡献度等评价因子；
三是顶层标准，强调"设计之力"，包含社会影响力、行业示范力等评价因子。

关于"设计知识产权保护"的声明

知识产权是智力劳动者对其成果依法享有的一种权利，随着设计在企业竞争中的重要性不断上升，依靠知识产权的维系设计创新领先的企业优势，尊重设计师的智力劳动并能与社会保持良好的互动，确立"中国制造"在世界的崭新面貌，保护设计知识产权已刻不容缓。中国作为世界知识产权组织的成员国，自加入时起至今，已经不断建立、完善并执行各项知识产权相关的法律法规，做出了巨大成绩，因此，中国设计智造大奖作为中国第一个综合性、国际性的工业设计"学院奖"，理应在此负有使命。在此，组委会郑重声明：

1. 中国设计智造大奖将秉持有关国际与中国知识产权保护条约的精神，在国家知识产权管理部门的支持下，尊重原创，努力为入选的获奖作品提供公开、公正的知识产权保护服务；

2. 我们承诺 DIA 所有的获奖作品必须均为原创，每届获奖作品的原创审议完全接受社会监督，有效期是终身的；

3. 大奖的执行机构"浙江现代智造促进中心"将成立专门的设计知识产权保护服务部门，对获奖作品在生产、销售、服务等所有环节中产生的违反知识产权保护有关法律的行为，协助申诉者开展工作，这项服务也是终身的；

4. 呼吁所有关心设计的公众一起共同努力，你们手中的消费选择权，是塑造设计未来的最重要力量，只有所有消费者摒弃山寨，支持原创，设计的知识产权保护才能真正实现。

中国设计智造大奖组委会办公室

ORIGIN

During the "Design Design" exhibition in October 2014, Comrade Li Qiang, together with the then president of China Academy of Art, Xu Jiang, put forward the idea of establishing an international industrial design award and decided that it would be hosted by China Academy of Art. On January 8, 2016, Comrade Yuan Jiajun officially announced the launch of DIA(Design Intelligence Award) .

DIA Overview

Design Intelligence Award (DIA) is a contemporary innovative design platform established in 2015 for the evaluation, promotion and collaboration of design. It is a new international award of industrial design aiming to remold the standards of contemporary design, a worldwide interdisciplinary competition encompassing art, technology and commerce which serves as the catalyst for the transformation of original ideas into materialized profit and the actualization of future innovation.

DIA Concepts

With "Intelligence of Humanity, Wisdom of Life, Fusion of Tech & Art, Brain of Industry" as the core values, DIA advocates that design return to its origin "Design Intelligence (Zhi Zao)". This is a design competition initiated by China and shared by the world. The essence of design is to put human life in the first place. The ancient lifestyle that "Living by coarse food, clothed in simple thread, drinking by pottery and sheltering near trees and streams" reflects the sustainable development conceptions of oriental material civilization, which contains the life wisdom and the notion of design. "Zhi" refers to not only the intelligence, wisdom, tactfulness, capability, etc., but also the awareness of methods of design and their correspondence to man, nature and the greater laws of universe. "Zao" refers to the design of forms, materials, settings and conceptions, which reflects ancient Chinese culture and grand design vision. In conclusion, design can be viewed as the integration of "Zhi" and "Zao", that is "Zhi Zao". From the perspective of Chinese wisdom, DIA reviews the modern design concepts that originated from western industrial civilization and proposes "Zhi Zao" to lead the development of the design industry in a new era.

Awards Evaluation System

There are three rounds of evaluation, namely, Preliminary Evaluation, Second Evaluation and Final Evaluation. In the Preliminary Evaluation, the images and text descriptions of your design works will be evaluated; in the Second Evaluation, the physical works and video introduction will be evaluated; and the Final Evaluation, the participants are required to participate in the oral defense on site. DIA Expert Bank consists of design masters and cross-discipline experts from all over the world, and the nomination experts and the evaluation experts will work independently.

Evaluation Criteria

In the context of the intelligent manufacturing era, the unique "DIA Evaluation System" includes three layers of criteria:

1. The fundamental layer emphasizes the "Principles of Design", covering functionality, aesthetics,technicality, user-experience and sustainability.

2. The advanced layer emphasizes the "Direction of Design", covering evaluation factors such as contribution to humanity, industry and the future.

3. The top layer emphasizes the "Impact of Design" covering evaluation factors such as social influence and exemplary role for the industry.

Declaration on "Design Intellectual Property Protection"

Intellectual Property Protection is a right enjoyed by intellectual laborers according to the law. It is imperative to strengthen Intellectual Property Protection to remain design innovation leading edge of enterprises, respect designers' intelligent labor and keep a good interaction with society, and establish a brand new look of "Made in China"in the world as design has played an increasingly important role in the enterprise competition. China, as a member of the World Intellectual Property Organization, since her accession, all Intellectual Property Protection related laws and regulations have been establishing, perfecting, implementing with great achievement. Therefore, Design Intelligence Award should undertake the responsibilities as the first comprehensive international "Academy Award" of industrial design in China. Hereby Design Intelligence Award Committee solemnly declares as below:

1. Design Intelligence Award will adhere to the spirit of the international and Chinese intellectual property protection treaties, with support of the State Intellectual Property Management Department, we respect originality and make great efforts to provide an open and fair intellectual property protection service for finalists and winners.

2. We are committed to make sure all award-winning works must be original, and the originality deliberation of each DIA winner will fully be subject to public supervision with lifelong validity.

3. "Zhejiang Modern Intelligence and Manufacturing Promotion Center"as the DIA executing organization will set up the special Design Intellectual Property Protection Department in dealing with violation against intellectual property rights protection relevant laws during all processes like award-winning works' production, sales and services. We will assist the applicants to carry out work, and this service will also last lifelong.

4.Appeal to the masses concerned about design work with joint efforts. Consumers' right of choice is the most important power to shape the future of design, only if all the consumers can reject cheap copy and support originality, can Design Intellectual Property Protection finally be realized.

Design Intelligence Award Committee Office

DIA 奖项设置

产业组	20项	460万元
金奖	2项	100万元 / 项
银奖	8项	20万元 / 项
铜奖	10项	10万元 / 项

概念组	9项	40万元
明日之星奖	2项	8万元 / 项
设计新锐奖	7项	3万元 / 项

佳作奖（约318项，实际数量以评审结果为准）

DIA Award System

Product Group	20 Awards	4,600,000RMB
Gold Award	2 Awards	1,000,000RMB/Award
Silver Award	8 Awards	200,000RMB/Award
Bronze Award	10 Awards	100,000RMB/Award

Concept Group	9 Awards	400,000RMB
Future Talents Award	2 Awards	80,000RMB/Award
Young Talents Award	7 Awards	30,000RMB/Award

Honorable Mention Award（Approximately 318 items, with the actual number subject to the evaluation results）

"金智塔"评价体系
"DIA Evaluation System"

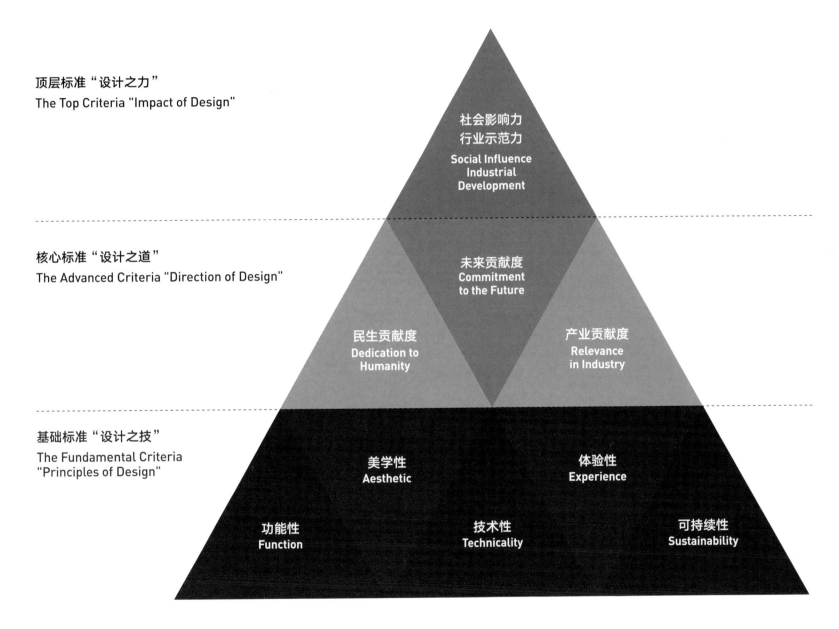

顶层标准"设计之力"
The Top Criteria "Impact of Design"

社会影响力
行业示范力
Social Influence
Industrial
Development

核心标准"设计之道"
The Advanced Criteria "Direction of Design"

未来贡献度
Commitment
to the Future

民生贡献度
Dedication to
Humanity

产业贡献度
Relevance
in Industry

基础标准"设计之技"
The Fundamental Criteria
"Principles of Design"

美学性
Aesthetic

体验性
Experience

功能性
Function

技术性
Technicality

可持续性
Sustainability

覆盖 70 多个国家和地区
Covering 70+ Countries and Regions

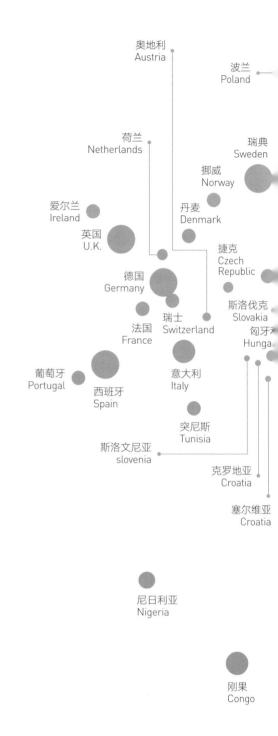

奥地利
Austria

波兰
Poland

荷兰
Netherlands

瑞典
Sweden

挪威
Norway

丹麦
Denmark

爱尔兰
Ireland

英国
U.K.

捷克
Czech
Republic

加拿大
Canada

德国
Germany

斯洛伐克
Slovakia

法国
France

瑞士
Switzerland

匈牙利
Hunga

美国
USA

葡萄牙
Portugal

西班牙
Spain

意大利
Italy

墨西哥
Mexico

突尼斯
Tunisia

斯洛文尼亚
slovenia

哥斯达黎加
Costa rica

克罗地亚
Croatia

哥伦比亚
Colombia

塞尔维亚
Croatia

尼日利亚
Nigeria

秘鲁
Peru

巴西
Brazil

刚果
Congo

智利
Chile

乌拉圭
Uruguay

阿根廷
Argentina

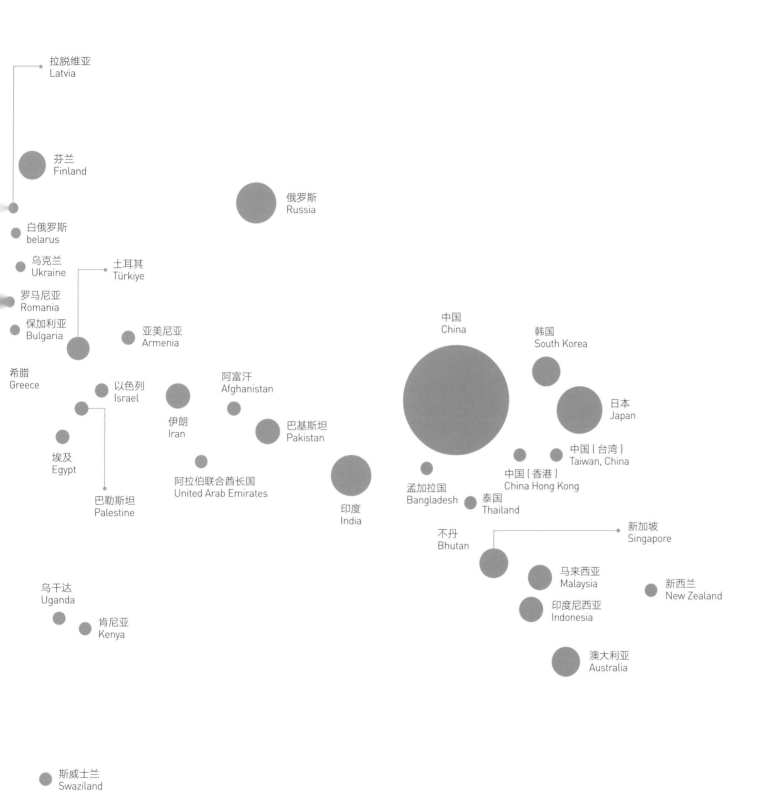

拉脱维亚
Latvia

芬兰
Finland

俄罗斯
Russia

白俄罗斯
belarus

乌克兰
Ukraine

土耳其
Türkiye

罗马尼亚
Romania

保加利亚
Bulgaria

亚美尼亚
Armenia

中国
China

韩国
South Korea

希腊
Greece

以色列
Israel

阿富汗
Afghanistan

日本
Japan

伊朗
Iran

巴基斯坦
Pakistan

埃及
Egypt

中国〔台湾〕
Taiwan, China

巴勒斯坦
Palestine

阿拉伯联合酋长国
United Arab Emirates

中国〔香港〕
China Hong Kong

孟加拉国
Bangladesh

泰国
Thailand

印度
India

不丹
Bhutan

新加坡
Singapore

乌干达
Uganda

马来西亚
Malaysia

新西兰
New Zealand

肯尼亚
Kenya

印度尼西亚
Indonesia

澳大利亚
Australia

斯威士兰
Swaziland

南非
South Africa

获奖作品来源分布
Distribution of Award-Winning Works by Country and Region in 2023

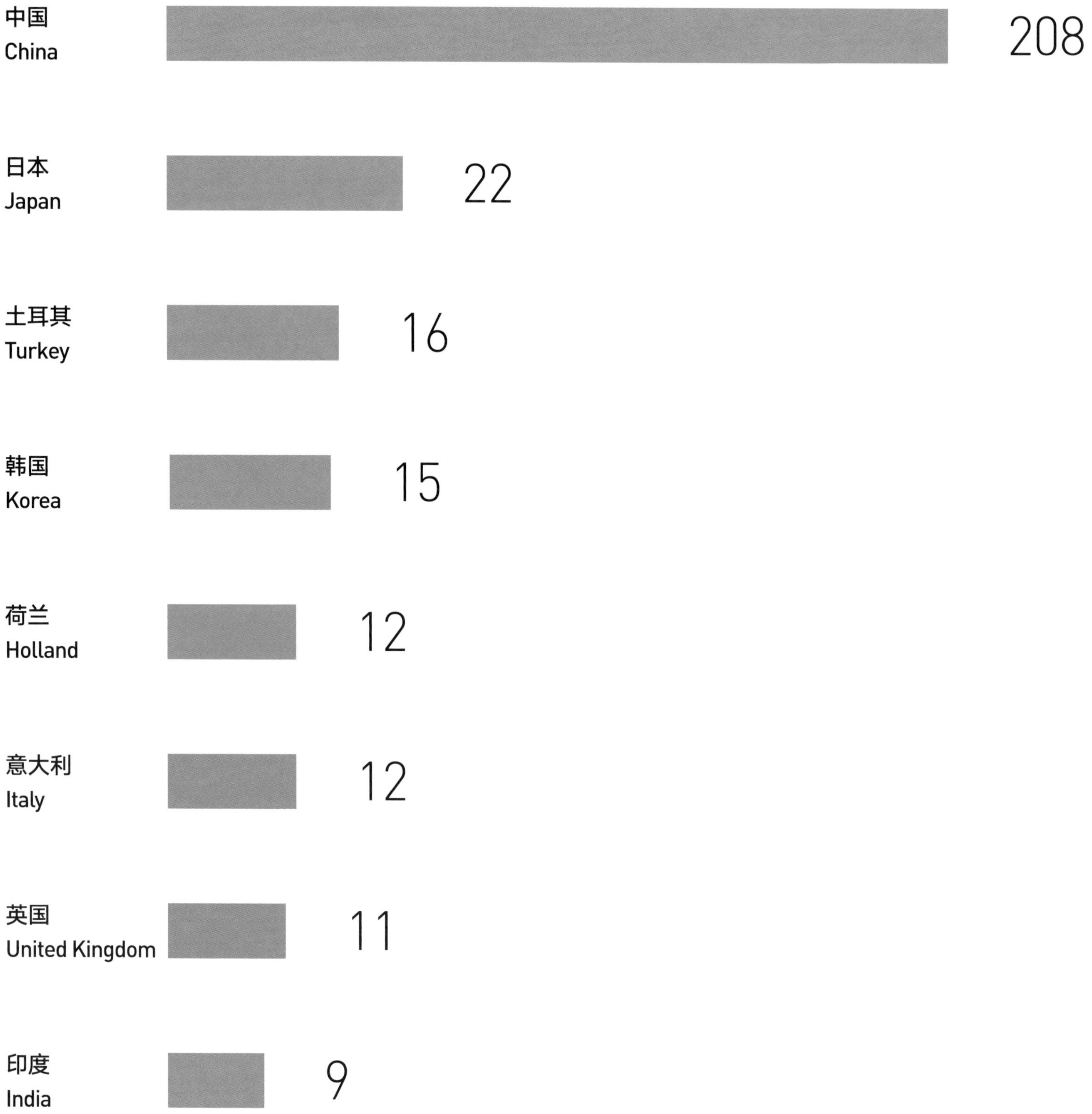

中国
China   208

日本
Japan   22

土耳其
Turkey   16

韩国
Korea   15

荷兰
Holland   12

意大利
Italy   12

英国
United Kingdom   11

印度
India   9

肯尼亚 Kenya	8	新加坡 Singapore	2
美国 America	8	中国（香港） Hong Kong,China	2
中国（台湾） Taiwan,China	6	尼日利亚 Nigeria	1
奥地利 Austria	5	斯洛伐克共和国 Slovak Republic	1
德国 Germany	4	乌干达 Uganda	1
法国 France	3	新西兰 New Zealand	1
墨西哥 Mexico	2	以色列 Israel	1
西班牙 Spain	2		

2022 DIA 视觉形象设计说明
VISUAL IDENTITY DESIGN INTERPRETATION

2022 DIA 视觉形象设计延续往届的视觉基因，以液态金属与变化的球体结构为基础元素共筑设计共同体 ——"DIA 星球"，金、银两种金属材质分别对应艺术智性与科技智能，体现了设计在两者的汇聚与碰撞中协同进化，生动诠释了"设计协同"的年度主题。

持续变化的液态生长方式，体现了 DIA 星球作为设计共同体，在艺术智性 + 科技智能的双 AI 赋能的助力下，如何探索可以融合的新状态；呈现了设计在生产、生活、生态不同维度的探索途径，在科技艺术背景下不断突破壁垒的融合过程；最终生成的球体结构 —— DIA 星球是设计师们在"协同可续"中不断地打破、重建后构筑起的新关系与新网络。

The visual genes of the previous DIA events have continued in the visual image design of 2022 DIA, with which we build a design community -"DIA Planet" with liquid metal and a changing sphere structure as the basic elements. Two metal materials-gold and silver-correspond to artistic intelligence and technological intelligence respectively, which reflects the coevolution of design in the convergence and collision of the two metals, and vividly interprets the annual theme of "Design Collaboration".

The continuously changing liquid growth mode reflects how DIA Planet, as a design community, explores a new state that can be integrated with the support of the dual AI empowerment: artistic intelligence + technological intelligence; It presents the exploration paths of design in different dimensions including production, life and ecology, and as well as the integration process of constantly breaking through the barriers in the context of technology and art; DIA Planet, the spherical structure, is a new relationship generated and a network built by designers after constantly breaking and rebuilding design in "Collaborative Sustainability".

2022 Design
Intelligence Award
中国设计智造大奖

赛事历程
COURSE OF THE EVENT

▼ 2022.04.09

第七届中国设计智造大奖启动全球征集，联合全球合作伙伴再次发起设计倡议，呼吁全球制造企业、设计公司、设计院校等组织，开启"协同可续"的探索。

The Call for Entry of 2022 DIA began, and DIA jointly launched a design initiative with global partners, calling on global manufacturing enterprises, design companies, design colleges and other organizations to start the exploration of "Collaborative Sustainability".

▼ 2022.07.15-08.05

第七届中国设计智造大奖初评采用线上评审的方式举行。全球44个国家和地区的8010件作品参加初评，由13位中国主赛区评委与意大利、荷兰、英国、法国、奥地利、日本、韩国、美国、加拿大、土耳其、墨西哥、肯尼亚等12个分赛区评委参与评审。初评评审后，共有840件作品成功入围并展开激烈角逐。

The 2022 DIA Preliminary Evaluation was held online. 8,010 entries from 44 countries and regions around the world participated in the Preliminary Evaluation, with 13 judges from China and judges from 12 overseas divisions including Italy, the Netherlands, the United Kingdom, France, Austria, Japan, South Korea, the United States, Canada, Turkey, Mexico and Kenya participating in the evaluation. After the Preliminary Evaluation, a total of 840 entries were shortlisted for the fierce competition.

世界设计之都大会 World Design Cities Conference 2022

▼ 2022.09.15-09.18

DIA 受邀亮相首届世界设计之都大会 (World Design Cities Conference 2022)，旨在通过展现设计的跨界融合趋势、创新驱动效应和泛在赋能作用，激发新的设计思考方式。

DIA was invited to appear in the first World Design Cities Conference 2022, which aimed to stimulate new ways of design thinking by showing the cross-border integration trend, innovation-driven effect and ubiquitous empowerment of design.

▼ 2022.10.22-10.30

DIA 再度亮相荷兰设计周，DIA 组委会秘书长王昀教授受邀为开幕式致辞。今年 DIA 特展的主题是"桥"，意在打破壁垒，建立一种新的融合关系。

DIA was back at Dutch Design Week again, and Professor Wang Yun, Secretary General of DIA Committee, was invited to deliver a speech at the opening ceremony. The theme of 2022 DIA special exhibition is "Bridge" , which means breaking down barriers between people, objects and the world to establish a new integrated relationship.

荷兰设计周 Dutch Design Week

▼ 2022.11.02

DIA 设计开放日对话邀请全球优秀设计师、专家学者、创投伙伴、产业园代表、DIA 新媒体新青年等联合展开了一场链接全球设计师的协同探索。

During the "2022 DIA Design Open Talk" , outstanding designers, experts and scholars, venture capital partners, representatives of industrial parks from all over the world, and DIA new media and new youth were invited to jointly launch a collaborative exploration that links global designers together.

▼ 2022.11.03

第七届中国设计智造大奖总决赛采用线上线下相结合的方式，海内外专家评审团通过线上视频会议同步参加。13位中外评委遵循 DIA "金智塔" 评价体系，通过实物评审、会议讨论两个环节，以严谨的态度综合票选后，共评选出355件获奖作品。

The 2022 DIA Final Evaluation was carried out online and offline, and DIA judges at home and abroad participated in the Final Evaluation simultaneously through online video conference. Following the "DIA Pyramid Evaluation System", 13 Chinese and foreign judges rigorously selected 355 winners of this year through comprehensive voting after going through two links: physical evaluation and meeting discussions.

总决赛现场 Final Evaluation

▼ 2023.03.11

2022DIA 设计之夜暨福布斯中国设计评选颁奖典礼是业界资深专家、优秀设计师、知名企业家的精英狂欢夜，聆听大咖们真知灼见的深度分享，欣赏具有跨界才华设计师的魅力展示，建立设计界高端价值链接。

The Award Ceremony of 2022 DIA Design Night & 2022 Forbes China Design Award Ceremony is a gala night for elites such as senior experts, outstanding designers and well-known entrepreneurs in the industry, where the participants listened to the remarkable insights from big names, appreciated the excellent works of talented crossover designers, and established valuable links with the design industry.

▼ 2023.03.11-03.12

中国设计智造协同创新国际学术研讨会由中国美术学院主办，以"设计智造·协同创新"为总主题，旨在探讨人工智能时代下设计的广泛转型。全球20个国家和地区的100余位专家参与线上线下研讨。

China Academy of Art organized the International Conference on Design Intelligence and Collaborative Innovation. With the overall theme of "Design Intelligence Creation and Collaborative Innovation", the Conference aimed at exploring the extensive transformation of design in the era of artificial intelligence. Over 100 experts from more than 20 countries and regions around the world participated in the discussion online and offline.

▼ 2023.03.12

文创设计智造教育部哲学社会科学实验室（D-AI）国际设计研究学会（IFDR）DIA 设计智造产业同盟（DIU）联席年会，围绕两大议题展开，一是依托 DIA 衍生发展建立的产学研三个平台，对于2023的工作重点进行规划；二是交流产业与高校、科技与市场、学研与应用之间的协同创新方式，共同探讨产学研相互促进与共同发展之道。

The Joint Annual Conference of Design-AI Lab, International Forum for Design Research (IFDR) and Design Intelligence Union (DIU) focused on two major issues. First, relying on the three platforms of producers, universities and researchers derived from and developed and established by DIA, the Conference made plans for the priorities of 2023; Second, participants made exchanges on collaborative innovation methods between industry and universities, technology and market, research and application, and jointly discussed the mutual promotion and common development of producers, universities and researchers.

颁奖典礼 DIA Award Ceremony

▼ **2023.03.12**

第七届中国设计智造大奖颁奖典礼在绍兴国际会展中心隆重举行。来自苏州翼博特智能科技有限公司的作品"便携式智能光伏清扫机器人"和来自华为技术有限公司的作品"盘古药物分子大模型智能加速药物研发"斩获金奖，浙江省副省长卢山为金奖得主颁奖。

The 2022 DIA Award Ceremony was grandly held in Shaoxing International Convention and Exhibition Center. The IFBOT X3 Solar Panel Cleaning Robot designed by Suzhou ifbot Intelligent Technology Co., Ltd. and the Pangu Drug Molecule Model: Accelerating Drug Discovery with AI from Huawei Technologies Co., Ltd. won the Gold Award. Lu Shan, Vice Governor of Zhejiang Province, presented awards to the Gold Award winners.

浙江省副省长卢山为2022DIA 金奖获奖者颁奖
Lu Shan, the Deputy Governor of Zhejiang Province is Presenting Awards to Winners

▼ 2023.03.12-03.19

数化·进化 —— 中国设计智造大展以"数化·进化"为主题，依托 DIA 综合创新平台资源，分为设计联合国、文化创新、生活智慧、产业装备、数字经济、2065未来图景、Design Will 讲堂、设计新青年等展区，特邀13家 DIA 国际工作站及12家海外分赛区负责人共同参与策划与组织，共展出优秀设计作品近四百件。

With the theme of "Digitization and Evolution", Digitization·Evolution Design Intelligence Exhibition, based on the resources of DIA comprehensive innovation platform, is divided into exhibition areas such as Design Union, Cultural Innovation, Life Wisdom, Industrial Equipment, Digital Economy, 2065 Future Landscape, Design Will Lecture, Design New Youth, etc. The heads of 13 DIA international workstations and 12 overseas divisions were invited to participate in planning and organizing the exhibition, and nearly 400 outstanding design works were exhibited during the exhibition.

设计大展
DIA Exhibition

初评评委
JUDGES OF PRELIMINARY EVALUATION

按姓氏首字母排序
Alphabetical Order

主赛区
Main Competition Area

常冰 Chang Bing

中国 CHINA

哪吒汽车副总裁
Vice President, NETA

方海 Fang Hai

中国 CHINA

芬兰阿尔托大学研究员、博导
Researcher, Doctoral supervisor,
Aalto University

郭春方 Guo Chunfang

中国 CHINA

吉林大学艺术学院院长
Dean, College of Arts, Jilin University

胡晓 Hu Xiao

中国 CHINA

IXDC 国际体验设计大会主席
President, International Experience Design
Conference

贾京生 Jia Jingsheng

中国 CHINA

清华大学美术学院长聘教授、博导
印染实验室主任
Professor and Doctoral Supervisor, Academy
of Arts & Design, Tsinghua University
Director, Printing and Dyeing Laboratory,
Tsinghua University

覃京燕 Qin Jingyan

中国 CHINA

北京科技大学人工智能研究院副院长
Deputy Dean, Institute of Artificial Intelligence,
University of Science and Technology Beijing

孙守迁 Sun Shouqian

中国 CHINA

浙江大学未来实验室主任
Director, Future Lab, Zhejiang University

王昀 Wang Yun

中国 CHINA

中国设计智造大奖组委会秘书长
Design Intelligence Award Comittee

魏洁 Wei Jie

中国 CHINA

江南大学设计学院院长
Dean, School of Design, Jiangnan University

杨学太 Yang Xuetai

中国 CHINA

华侨大学工业设计研究院院长
国家陶瓷行业工业设计研究院院长
Dean, Institute of Industrial Design, Huaqiao University
President, National Institute of Ceramic Industry Design

张明 Zhang Ming

中国 CHINA

南京艺术学院工业设计学院院长
Dean, School of Industrial Design, Nanjing University of the Arts

张展 Zhang Zhan

中国 CHINA

上海设计之都促进中心理事长
Director, Shanghai Promotion Center for City of Design

周立钢 Zhou Ligang

中国 CHINA

杭州市工业设计协会会长
杭州博乐工业设计股份有限公司创始人
President, Hangzhou Industrial Design Association
Founder, Hangzhou Bole Industrial Design Co., Ltd.

日本分赛区
Japan Division

Kota Nezu

日本 JAPAN

znug design 公司 CEO
CEO, znug design, inc.

Nobuyoshi Yamazaki

日本 JAPAN

东京艺术大学副教授
Associate Professor, Tokyo University of the Arts

Shiro Aoki

日本 JAPAN

国际设计研究学会理事
Board Member, International Forum for Design Research

英国分赛区
UK Division

Amanda Broderick

英国 UNITED KINGDOM

教授，东伦敦大学校长，伦敦高校联盟主席
Professor, Vice-Chancellor & President, University of East London
Chair, London Higher

Jeff Cao

英国 UNITED KINGDOM

Aspire Ventures 董事
Director, Aspire Ventures Ltd

John Mathers

英国 UNITED KINGDOM

英国设计基金会主席
Chair, British Design Fund

Mark Hedley

英国 UNITED KINGDOM

英中贸易协会知识经济总监
Director Knowledge Economy, China-Britain Business Council

Russ Shaw

英国 UNITED KINGDOM

伦敦科技促进组织、全球科技倡导组织创始人
Founder, Tech London Advocates & Global Tech Advocates

法国分赛区
France Division

Fabien Grégoire

法国 FRANCE

创新品牌设计顾问
Innovation, Brand & Design Consultant

Jean Christophe NAOUR

法国 FRANCE

三星首席 UX/UI 设计师
Principal UX/UI Designer, Samsung Electronics

Stéphane Thirouin

法国 FRANCE

Williwaw Fan 公司 CCO
Chief Camelot Officer, Williwaw Fan

意大利分赛区
Italy Division

Cristian Confalonieri

意大利 ITALY

Studiolabo 联合创始人、创意总监
Co-founder & Creative Director,
Studiolabo

Luca Fois

意大利 ITALY

米兰理工大学设计学院教授
Professor, Design School, Politecnico
di Milano

Michele Brunello

意大利 ITALY

DONTSTOP Architettura 创始人
Founder, DONTSTOP Architettura

荷兰分赛区
Netherlands Division

Jun Hu

荷兰 NETHERLANDS

欧中设计促进会主席
Chairman, Stichting voor Design Promotion in
Europa en China

Loe Feijs

荷兰 NETHERLANDS

埃因霍温理工大学名誉教授
Emeritus Professor, Eindhoven University of
Technology

Marina Toeters

荷兰 NETHERLANDS

Fashion Tech Farm 联合创始人
Co-founder, Fashion Tech Farm

韩国分赛区
South Korea Division

Byung-wook Chin

韩国 SOUTH KOREA

汉城大学教授
Professor, Hansung University

Hyunsun Kim

韩国 SOUTH KOREA

KFDA 主席
President, Korean Federation of Design
Associations

Kang-Heui Cha

韩国 SOUTH KOREA

弘益大学教授
Professor, Hongik University

加拿大分赛区
Canada Division

Balkrishna Mahajan

加拿大 CANADA

Ticket Design 联合创始人
Co-founder, Ticket Design

James Vanderpant

加拿大 CANADA

Çava 联合创始人、设计主管
Co-founder & Head of Design, Çava

Sarang Sheth

加拿大 CANADA

Yanko Design 主编
Editor in Chief, Yanko Design

美国分赛区
US Division

Allan Chochinov

美国 UNITED STATES

Core77合伙人
Partner, Core77

Beau Oyler

美国 UNITED STATES

Enlisted Design 公司 CEO、首席设计师
CEO & Chief Design Officer, Enlisted Design

Brian Roderman

美国 UNITED STATES

IN2 Innovation 主席兼首席创新官
President & Chief Innovation Officer, IN2 Innovation

奥地利分赛区
Austria Division

Anna Maislinger

奥地利 AUSTRIA

IN PRETTY GOOD SHAPE 联合创始人
Co-founder, IN PRETTY GOOD SHAPE

Florian Halm

奥地利 AUSTRIA

信息设计协会（IDA）董事会成员
Board Member, Information Design Association

Mathias Schasché

奥地利 AUSTRIA

RDD design network 工业设计师
Industrial Designer, RDD design network

墨西哥分赛区
Mexico Division

Francisco Javier Santín Reyna

墨西哥 MEXICO

墨西哥州立自治大学建筑与设计学院专业评估主任
Head of the Professional Evaluation Department, Architecture and Design Faculty, Autonomous University of the State of Mexico

María del Consuelo Espinosa Hernández

墨西哥 MEXICO

墨西哥州立自治大学建筑与设计学院工业设计系主任
Head of Industrial Design Program, Faculty of Architecture and Design, Autonomous University of the State of Mexico

Martha Patricia Zarza Delgado

墨西哥 MEXICO

墨西哥州立自治大学高等研究院秘书长
Secretary of Research and Advanced Studies, Autonomous University of the State of Mexico

肯尼亚分赛区
Kenya Division

Adrian Jankowiak

肯尼亚 KENYA

内罗毕设计周创始人
Founder, Nairobi Design Week

Fenoson Zafimahova

肯尼亚 KENYA

首席概念设计师
Lead Concept Designer

Mugendi K. M' Rithaa

肯尼亚 KENYA

马查科斯大学教授
Professor, Machakos University

土耳其分赛区
Turkey Division

Fatih Mintas

土耳其 TURKEY

Silverline Appliances 产品经理
Product Manager, Silverline Appliances

Ilgaz Kuruyazici

土耳其 TURKEY

FUTURE-IST DESIGN & CONSULTANC 公司
创始人
Founder, FUTURE-IST DESIGN & CONSUL-
TANC

Sertac Ersayin

土耳其 TURKEY

WDO 董事会成员，土耳其工业设计师协会
主席
Board Member, World Design Organization
President, Industrial Designers Society of
Turkey

总决赛评委
JUDGES OF FINAL EVALUATION

按姓氏首字母排序
Alphabetical Order

Nicolas Cinguino

法国 FRANCE

Blank Group 合伙人
Co-founder, Blank Group

Sertac Ersayin

土耳其 TURKEY

WDO 董事会成员，土耳其工业设计师协会主席
Board Member, World Design Organization
President, Industrial Designers Society of Turkey

郭春方 Guo Chunfang

中国 CHINA

吉林大学艺术学院院长
Dean, College of Arts, Jilin University

何人可 He Renke

中国 CHINA

湖南大学学术委员会主席
Academic Chairman, Hunan University

James B. Hope

英国 UNITED KINGDOM

中国美术学院交通工具设计研究所所长
Director, Transportation Design Institute,
China Academy of Art

李超德 Li Chaode

中国 CHINA

苏州大学博物馆馆长
Director, Soochow University Museum

孙守迁 Sun Shouqian

中国 CHINA

浙江大学未来实验室主任
Director, Future Lab, Zhejiang University

Kostas Terzidis

美国 UNITED STATES

同济大学设计创意学院教授，尚想实验室主任
Professor, College of Design and Innovation,
Tongji University
Director, ShangXiang Lab

宗福季 TUSNG Fugee

中国（香港）HONG KONG, CHINA

香港科技大学（广州）信息枢纽署理院长
Acting Dean, Information Hub, The Hong
Kong University of Science and Technology
(Guangzhou)

王昀 Wang Yun

中国 CHINA

中国设计智造大奖组委会秘书长
Secretary-general of Design Intelligence
Award Committee

张凌浩 Zhang Linghao

中国 CHINA

南京艺术学院校长
President, Nanjing University of the Arts

赵健 Zhao Jian

中国 CHINA

上海大学博导，澳门科技大学博导
Doctoral Supervisor, Shanghai University
Doctoral Supervisor, Macau University of
Science and Technology

赵超 Zhao Chao

中国 CHINA

清华大学美术学院副院长
Deputy Dean, Academy of Arts & Design,
Tsinghua University

关于 DIU
ABOUT DIU

DIU 发起大会

DIU 国际设计智造联盟 (Design Intelligence Union,DIU) 发起会议于 2023 年 3 月 12 日召开，中国美术学院院长高世名，中国美术学院副院长韩绪，中国设计智造大奖组委会秘书长、国际设计智造联盟（DIU）召集人卢涛，以及网易高级副总裁、网易雷火游戏事业群总裁胡志鹏，摩尔线程、摩尔学院院长李丰，华为 UCD 中心设计总监王守玉，鱼跃医疗设计总监华昊等 DIU 发起单位的代表出席会议并发言。

立足新时代高质量创新发展观，国际设计智造联盟（DIU）携手国际设计师组织、国内外艺术院校以及数字科技与先进制造企业，共同发起联盟倡议，以"开放、创新、协作、共享"为理念，以"聚焦产业、面向国际、塑造未来"为宗旨，以艺科商融合为导向，以智能设计集成创新为手段，吸纳国内外多元主体广泛参与"政产学研用"协同创新的探索与实践，打造具有全球影响力的交流合作平台，构建设计创新与智造产业深度融合的跨行业网络，共同推动产业升级和社会进步。

DIU Launch Meeting

The DIU launching meeting was held on March 12, 2023. Gao Shiming, President of China Academy of Art, Han Xu, Vice President of China Academy of Art, Lu Tao, Secretary General of the Design Intelligence Award (DIA) Committee and Convener of Design Intelligence Union (DIU), Hu Zhipeng, Senior Vice President of Netease and President of NetEase Thunder Fire Studio, Li Feng, President of Moore College, Moore Threads, Wang Shouyu, Design Director of Huawei UCD Center, Hua Hao, Design Director of Yuyue Medical Equipment & Supply Co., Ltd., and other representatives of DIU initiators attended the meeting and delivered speeches.

Based on the high-quality innovation development concept in the new era, the Design Intelligence Union (DIU), together with international designer organizations, art colleges at home and abroad, and digital technology and advanced manufacturing enterprises, jointly launched the DIU initiative. Taking "openness, innovation, collaboration and sharing" as the concept, "focusing on industry, facing the world and shaping the future" as the goal, integration of arts, science and business as the guide and the design integration and innovation of intelligent design as the means, DIU aims at attracting domestic and foreign diverse subjects to participate in the exploration and practice of collaborative innovation of" government, industry, university, research and application" and creating a platform for exchanges and cooperation with global influence, building a cross-industry network with deep integration of design innovation and intelligent manufacturing industries, and jointly promoting industrial upgrading and social progress.

年会现场 Annual Conference

DIA 及 DIU 秘书长卢涛发言
Speech by Lu Tao, Secretary-General of DIA and DIU

网易高级副总裁 网易雷火游戏事业群总裁胡志鹏发言
Speech by Hu Zhipeng, Senior Vice President of Netease and President of NetEase Thunder Fire Studio

摩尔线程、摩尔学院院长李丰发言
Speech by Li Feng, President of Moore College, Moore Thread

华为 UCD 中心设计总监王守玉发言
Speech by Wang Shouyu, Design Director of Huawei UCD Center

鱼跃医疗设计总监华昊发言
Speech by Hua Hao, Design Director of Jiangsu Yuyue Medical Equipment & Supply Co., Ltd.

颁发 DIU 成员单位证书
Certificate Presentation to DIU Member Units

高世名院长、韩绪副院长为 DIU 联盟成员颁发证书
The President Gao Shiming and Vice-president Han Xu are presenting certificates to DIU members

DIU 联合发起单位
The DIU Initiators

DIU 国际设计智造联盟发起单位共 28 家，覆盖跨行业平台、智能装备、智慧医疗、智慧出行、智慧家居五大领域的头部企业。

The Design Intelligence Union (DIU) is co-initiated by a total of 28 leading enterprises spanning across five major domains: cross-industry platforms, intelligent equipment, smart healthcare, smart transport, and smart home.

1. 中国美术学院 China Academy of Art

中国美术学院前身为1928年由蔡元培、林风眠先生创建的国立艺术院。95年来，学校以强烈的"名校意识""自主意识"打造国美模式，培养了一大批德艺双馨的优秀人才，对全国艺术教育起到示范引领作用。

The China Academy of Art, formerly known as the National Academy of Arts, was founded in 1928 by Mr. Cai Yuanpei and Mr. Lin Fengmian. Over its 95-year history, the university has developed a strong sense of excellence and independence, creating the "China Academy of Arts model." It has nurtured a significant number of outstanding talents who excel both in moral integrity and professional skills. The China Academy of Art plays a leading and exemplary role in art education in China.

2. 科大讯飞股份有限公司 IFLYTEK Incorporated Company

科大讯飞股份有限公司成立于1999年，是亚太地区知名的智能语音和人工智能上市企业。自成立以来，一直从事智能语音、自然语言理解、计算机视觉等核心技术研究并保持了国际前沿技术水平；积极推动人工智能产品和行业应用落地，致力让机器"能听会说，能理解会思考，用人工智能建设美好世界"。2008年，公司在深圳证券交易所挂牌上市（股票代码：002230）。

Founded in 1999, iFLYTEK is a well-known listed company specializing in intelligent speech and artificial intelligence technologies in the Asia-Pacific region. Since its establishment, the company is devoted to cornerstone research in technologies including intelligent speech, natural language understanding and computer vision and has maintained a world-leading position in those domains. The company actively promotes the development of AI products and their sector-based applications, with visions of enabling machines to listen and speak, understand and think, and creating a better world through artificial intelligence. In 2008, iFLYTEK became a publicly traded company on the Shenzhen Stock Exchange (Stock Code: 002230).

3. 杭州涂鸦信息技术有限公司 Hangzhou Tuya Information Technology Inc.

杭州涂鸦信息技术有限公司是一家致力于让生活更智能的领先技术公司，提供能够智连万物的云平台，打造互联互通的开发标准，连接品牌、OEM 厂商、开发者、零售商和各行业的智能化需求，涂鸦的解决方案赋能并提升合作伙伴和客户的产品价值，同时通过技术应用使消费者的生活更加便利，涂鸦智能的智慧商业 SaaS 为丰富的垂直行业提供智能解决方案。

Hangzhou Tuya Information Technology Inc. is a leading technology company dedicated to making life smarter. Tuya provides a cloud platform that can connect everything intelligently, creating an interconnected development standard. It serves as the bridge between brands, OEM manufacturers, developers, retailers, and various industries' smart needs. Tuya's solutions empower and enhance the product value of partners and customers. At the same time, it makes consumers' lives more convenient through technological applications. Tuya Smart's intelligent business SaaS offers intelligent solutions for a wide range of vertical industries.

4. 网易（杭州）网络有限公司 NetEase Inc.

网易（杭州）网络有限公司成立于1997年，是中国领先的互联网技术公司，也是中国主要门户网站。网易公司推出了门户网站、在线游戏、电子邮箱、在线教育、电子商务、在线音乐等多种服务。

NetEase Inc. founded in 1997, is a leading Chinese internet technology company and a major online portal in China. NetEase offers a wide range of services, including online portals, online games, email services, online education, e-commerce and online music.

5. 华人运通 Human Horizons

华人运通是由丁磊于2017年创立，以智能汽车、智捷交通、智慧城市"三智"为战略布局，是一家专注于未来智能交通产业的创新型出行科技公司。旗下品牌为中国高端智能电动车品牌 —— 高合汽车，主要产品包括双旗舰车型高合 HiPhi X、高合 HiPhi Z 以及主力舰车型高合 HiPhi Y。

Human Horizons was founded by Ding Lei in 2017, with a strategic focus on "Three Intelligences" – smart vehicles, intelligent transportation, and smart cities. It is an innovative travel technology company specializing in the future intelligent transportation industry. Under its umbrella, there is a Chinese high-end smart electric vehicle brand called "Highway" (HiPhi Auto), with its primary products including the flagship models HiPhi X and HiPhi Z, as well as the key model HiPhi Y.

6. 杭州灵伴科技有限公司 Hangzhou Lingban Technology Co., Ltd.

杭州灵伴科技有限公司创立于2014年，是一家专注于人机交互技术的产品平台公司，深耕5G+AI+AR 领域的软硬件产品开发，通过语音识别、自然语言处理、计算机视觉、光学显示、芯片平台、硬件设计等多领域研究，将前沿的 AI 和 AR 技术与行业应用相结合，为不同垂直领域的客户提供全栈式解决方案，打造智能时代的超级工人，有效提升用户体验、助力企业增效。

Hangzhou Lingban Technology Co., Ltd. was founded in 2014 and is a product platform company specializing in human-computer interaction technology. The company deeply engages in the development of software and hardware products in the 5G+AI+AR fields. By conducting research in multiple areas such as speech recognition, natural language processing, computer vision, optical displays, chip platforms, and hardware design, it integrates cutting-edge AI and AR technologies with industry applications. Lingban Technology provides full-stack solutions to clients in various verticals, aiming to create superworkers for the intelligent era, effectively enhancing user experiences, and boosting enterprise efficiency.

7. 合肥联宝信息技术有限公司 Hefei LCFC Information Technology Co., Ltd.

合肥联宝电子科技有限公司（以下简称联宝科技）成立于2011年，是联想（全球）最大的智能计算设备研发和制造基地，2023年入选世界"灯塔工厂"，属于国家评定的智能制造试点示范、工业设计中心、绿色工厂、国家知识产权示范企业和智能制造标杆企业。

Hefei LCFC Information Technology Co., Ltd.（hereinafter referred to as Lianbao Technology）was established in 2011. It is the largest global research and manufacturing base for intelligent computing devices under Lenovo. In 2023, it was selected as a "Lighthouse Factory" in the world. It is recognized as a national intelligent manufacturing pilot demonstration, an industrial design center, a green factory, a national intellectual property demonstration enterprise, and a benchmark enterprise in intelligent manufacturing.

8. 西子电梯科技有限公司 iLIFT | XIZI Ltd.

西子电梯科技有限公司是西子集团旗下电梯企业，是一家集研发、制造、销售、安装及售后服务为一体的综合型电梯制造企业。西子电梯多年来秉持"以创新驱动发展，用智慧联合创造"理念，不断提高科研、创新能力，已获得发明、实用新型、外观设计等多项国家专利。

XIZI Elevator Technology Co.,Ltd. is an elevator company under the XIZI Group. It is a comprehensive elevator manufacturing enterprise that integrates research and development, manufacturing, sales, installation, and after-sales service. Over the years, Xi Zi Elevator has adhered to the philosophy of "driving development through innovation and creating with wisdom in unity." They have continuously improved their research and innovation capabilities, resulting in the acquisition of numerous national patents, including inventions, utility models, and exterior designs.

9. 索尼（中国）有限公司 Sony（China）Co.,Ltd.

索尼（中国）有限公司上海分公司成立于1997年9月16日，注册地位于上海市湖滨路222号八楼，法定代表人为吉田武司。经营范围包括在批准的索尼（中国）有限公司的经营范围内，从事业务活动；出版物经营。

Sony（China）Co., Ltd. Shanghai Branch was established on September 16, 1997, with its registered address located on the 8th floor at 222 Hubin Road, Shanghai. The legal representative is YOSHIDA TAKESHI. Its business scope includes engaging in business activities within the approved scope of Sony（China）Co., Ltd., as well as publication operations.

10. 深圳华大智造科技股份有限公司 MGI Technology Incorporated Company

深圳华大智造科技股份有限公司（简称华大智造）成立于2016年，秉承"创新智造引领生命科技"的理念，致力于成为生命科技核心工具缔造者，专注于生命科学与生物技术领域，以仪器设备、试剂耗材等相关产品的研发、生产和销售为主要业务，为精准医疗、精准农业和精准健康等行业提供实时、全景、全生命周期的生命数字化设备和系统。

MGI Technology Incorporated Company(referred to as Huada Zhizao) was founded in 2016. Guided by the concept of "Innovative Manufacturing Leading Life Sciences," the company is dedicated to becoming a core tool provider in the field of life sciences and biotechnology. Their primary focus is on the research, development, production, and sale of instruments, equipment, reagents, and consumables related to life science and biotechnology. They aim to provide real-time, comprehensive, and full-lifecycle digital equipment and systems for industries such as precision medicine, precision agriculture, and precision health.

11. 方太集团 FOTILE Group

方太集团创建于1996年。作为一家以智能厨电为核心业务的幸福生活解决方案提供商，方太长期致力于为人们提供高品质的产品和服务，打造健康环保、有品位、有文化的生活方式，让千万家庭享受更加幸福安心的生活。FOTILE 方太专注于高端厨电的研发与生产，现拥有集成烹饪中心、吸油烟机、水槽洗碗机等多条产品线。

FOTILE Group was founded in 1996. As a provider of solutions for a joyful life centered around intelligent kitchen appliances, FOTILE has long been committed to offering people high-quality products and services. They aim to create a lifestyle that is healthy, environmentally friendly, cultured, and full of taste, allowing millions of families to enjoy a happier and more secure life. FOTILE specializes in the research and production of high-end kitchen appliances and currently offers multiple product lines, including integrated cooking centers, range hoods, and sink dishwashers.

12. 浙江苏泊尔股份有限公司 Zhejiang SUPOR Limited by share Ltd.

浙江苏泊尔股份有限公司是中国最大、全球第二的炊具研发制造商，中国厨房小家电领先品牌。苏泊尔于1994年在台州创立，2002年总部迁至杭州，在杭州、台州、绍兴、武汉和越南建立了5大研发制造基地，拥有10000多名员工。苏泊尔是中国炊具行业首家上市公司（股票代码002032）。

Zhejiang Supor Limited by share Ltd. is China's largest and the world's second-largest cookware research and manufacturing company, and a leading brand in China's kitchen small appliances industry. Supor was founded in 1994 in Taizhou and relocated its headquarters to Hangzhou in 2002. The company has established five major research and manufacturing bases in Hangzhou, Taizhou, Shaoxing, Wuhan, and Vietnam, with more than 10,000 employees. Supor was the first publicly listed company in China's cookware industry, with the stock code 002032.

13. 中车株洲电力机车有限公司 CRRC Zhuzhou Locomotive Co., Ltd.

中车株洲电力机车有限公司是中国中车股份有限公司旗下龙头企业，地处南方工业重镇和交通枢纽湖南省株洲市，公司前身为株洲电力机车厂，始建于1936年，是中国轨道电力牵引装备主要研制生产基地和城轨交通设备国产化定点企业，享有"中国电力机车之都"的美誉，也是国内唯一的电力机车整车出口企业。

CRRC Zhuzhou Electric Locomotive Co., Ltd. is a leading subsidiary of China CNR Corporation Limited (CRRC), located in Zhuzhou City, Hunan Province, a prominent industrial and transportation hub in southern China. The company's predecessor was Zhuzhou Electric Locomotive Factory, founded in 1936. It serves as a major research and production base for rail electrification equipment in China and a designated enterprise for the localization of urban rail transportation equipment.Zhuzhou Electric Locomotive Co., Ltd. is renowned as the "Capital of China's Electric Locomotives" and is the only domestic manufacturer of electric locomotives for whole vehicle exports.

14. 江苏鱼跃医疗设备股份有限公司 Jiangsu Yuwell Medical Equipment & Supply Co., Ltd.

江苏鱼跃医疗设备股份有限公司，中国A股上市公司（股票代码：002223），成立于1998年10月22日，集团总部设立在中国上海，拥有位于德国、意大利、中国北京等12大研发中心和9大制造中心，形成了完整的全球研发、生产、营销、服务网络，覆盖海外131个国家和地区。

Jiangsu Yuwell Medical Equipment & Supply Co., Ltd. is a publicly traded A-share company in China（Stock Code: 002223）. It was founded on October 22, 1998. The group's headquarters are located in Shanghai, China. Yuyue Medical has established 12 major research and development centers in countries such as Germany, Italy, and Beijing, China along with 9 major manufacturing centers. This has allowed them to create a comprehensive global network encompassing research, production, marketing, and service, reaching 131 countries and regions worldwide.

15. 添可智能科技有限公司 Tineco Inteligent Technology Co., Ltd.

添可倡导"生活白科技 居家小确幸"，以"以智能科技创造梦想生活"为品牌使命，致力于以科技为核心，创造无限居家可能，实现美好生活的梦想。成立以来，添可在智能家居清洁类电器领域不断探索升级，同时将智能生活电器延伸至个人护理、烹饪料理和健康生活品类，打造高端智能产品家族。

TINECO advocates "Smart Technology for a Better Life" and promotes "Little Joys at Home." With the brand mission of "Creating Dream Lives with Intelligent Technology," TINECO is committed to using technology as the core to create limitless possibilities for home life, realizing the dream of a better life. Since its establishment, TINECO has continually explored and upgraded its products in the field of smart home cleaning appliances. Additionally, it has extended its range of smart lifestyle appliances to personal care, cooking, and health products, aiming to build a family of high-end smart products.

16. 广州纽得赛生物科技有限公司 Guangzhou New-Design Biotechnology Co.,Ltd.

广州纽得赛生物科技有限公司是一家专注于家用医疗保健器械行业的集设计、研发、生产、销售于一体的高新技术企业。公司内部拥有专业的自主研发团队，致力于新兴产品的开发与设计，同时获得70多项国家专利。先后专注在"柔性发热材料""中医保健器械"等几个细分领域，进行了深度产品线布局，成功运作了"科爱元素""TECH LOVE"等品牌。

Guangzhou New-Design Biotechnology Co., Ltd. is a high-tech enterprise specializing in the design, research and development, production, and sales of home medical and healthcare devices. The company has an internal professional research and development team dedicated to the development and design of emerging products, and it has obtained more than 70 national patents.Newde has focused on several niche areas, including "flexible heating materials" and "traditional Chinese health care devices," and has made deep product line layouts in these fields. The company has successfully operated brands such as "科爱元素"（the element of science and love） and "TECH LOVE."

17. 深圳市科曼医疗设备有限公司 Shenzhen Comen Medical Instruments Co.,Ltd.

深圳市科曼医疗设备有限公司成立于2002年，是一家从事电生理监护、心电诊断、超声母婴监护、呼吸麻醉、婴儿保育、手术室设备研发的医疗器械公司。

Shenzhen Comen Medical Instruments Co., Ltd. was established in 2002 and is a medical equipment company specializing in the research and development of electrophysiological monitoring, electrocardiographic diagnosis, ultrasound maternal and child monitoring, respiratory anesthesia, infant care, and operating room equipment.

18. 珠海格力电器股份有限公司 Gree Electric Appliances Inc. of Zhuhai

珠海格力电器股份有限公司成立于1991年，1996年11月在深交所挂牌上市。公司成立初期，主要依靠组装生产家用空调，现已发展成为多元化、科技型的全球工业制造集团，产业覆盖家用消费品和工业装备两大领域，产品远销190多个国家和地区。

Gree Electric Appliances Inc. of Zhuhai was founded in 1991 and went public on the Shenzhen Stock Exchange in November 1996. In its early days, the company primarily focused on assembling household air conditioners. Since then, it has evolved into a diversified and technology-driven global industrial manufacturing group with operations in two main sectors: household consumer products and industrial equipment. Its products are exported to more than 190 countries and regions worldwide.

19. 亿咖通科技 ECARX

亿咖通科技是一家汽车智能化科技公司。企业以"让智慧出行驱动美好生活"为愿景，以"加速汽车智能化，创建人车新关系"为使命，致力于持续打造行业领先的智能网联生态开放平台，全面为车企赋能，创造更智能、更安全的出行体验。

ECARX is an automotive intelligent technology company. With the vision of "Enabling a Better Life through Smart Mobility" and the mission to "Accelerate Automotive Intelligence and Forge a New Relationship Between People and Cars," the company is dedicated to continuously building an industry-leading intelligent connected vehicle ecosystem open platform. Their goal is to empower automotive enterprises comprehensively, creating a smarter and safer travel experience.

20. 中电云计算技术有限公司 CECloud Computing Technology Co., Ltd.

中国电子云是中国信创云引领者，是中国电子自主计算产业体系的核心组成部分，坚定走自主技术创新的道路。秉承云数融合、市场牵引、商业成功以及从跟随到超越的产品体系理念，在数字基础设施建设运营、数据资源体系规划建设、数字技术的创新应用等领域全面布局，深度参与国家重大工程，服务政府及关键行业客户数字化转型。

China Electronic Cloud is the leader of China's Xinchuanchun cloud, is the core component of China's electronic autonomous computing industrial system, and firmly takes the road of independent technological innovation. Adhering to the concept of cloud data integration, market traction, commercial success and product system from following to surpassing, the company has a comprehensive layout in the fields of digital infrastructure construction and operation, data resource system planning and construction, and digital technology innovation and application, deeply participates in major national projects, and serves the digital transformation of government and key industry customers.

21. 新华医疗（智慧医疗）SHINVA Medical Instrument Co.,Ltd.

新华医疗成立于1943年，是我党我军创建的第一家医疗器械生产企业，2002年9月，在上海证券交易所上市（股票代码：600587），主营医疗器械、制药装备、医疗服务、医疗商贸四大业务板块，总部位于山东省淄博市。

SHINVA Medical Instrument Co.,Ltd. established in 1943, is the first medical equipment manufacturing enterprise created by the Chinese Communist Party and the People's Liberation Army. In September 2002, it went public on the Shanghai Stock Exchange (stock code: 600587). The company's main business segments include medical equipment, pharmaceutical equipment, medical services, and medical trade. Its headquarters are located in Zibo City, Shandong Province.

22. 三一集团有限公司 SANY Group Co., Ltd.

三一集团有限公司始创于1989年。自成立以来，秉持"创建一流企业，造就一流人才，做出一流贡献"的企业愿景，打造了业内知名的"三一"品牌。三一的使命是"品质改变世界"，即以极高品质的产品和服务改变中国产品的世界形象。目前三一正在实施三大战略：全球化，数智化，低碳化。

SANY Group Co., Ltd. was founded in 1989. Since its establishment, it has adhered to the corporate vision of "Creating a First-Class Enterprise, Cultivating First-Class Talent, and Making a First-Class Contribution," and has built the well-known "Sany" brand in the industry. Sany's mission is to "Change the World through Quality," aiming to improve the global image of Chinese products by providing products and services of extremely high quality. Currently, Sany is implementing three major strategies: globalization, digitalization, and low-carbonization.

23. 美的 - 美智纵横科技有限责任公司 Midea Robozone Technology Co.,Ltd.

美智纵横科技有限责任公司是美的集团旗下子公司，是一家聚焦人工智能技术、智能机器人产品和智慧解决方案的高科技公司。公司旗下拥有覆盖广泛消费人群的多品牌组合，众多品牌家喻户晓：Midea,Eureka, Toshiba 等。总部设于苏州，在美国硅谷，中国上海、深圳设有研发中心，在人工智能、软件算法、导航识别技术、光学声学研究、电子工程、机械结构设计与供应链管理等多领域，美智纵横拥有丰富的创新和实践经验。

Midea Robozone Technology Co., Ltd. is a subsidiary of the Midea Group, a high-tech company specializing in artificial intelligence technology, intelligent robot products, and smart solutions. The company owns a diverse portfolio of brands that cater to a wide range of consumers, with many well-known brands such as Midea, Eureka, Toshiba, and more. Headquartered in Suzhou, Mozztech has research and development centers in Silicon Valley, the United States, Shanghai, and Shenzhen, China. Across various fields, including artificial intelligence, software algorithms, navigation and recognition technologies, optical and acoustic research, electronic engineering, mechanical structure design, and supply chain management, Mozztech has accumulated rich experience in innovation and practical applications.

24. 浙江泰普森实业集团有限公司 Zhejiang Tapson Industrial Group Co., Ltd.

浙江泰普森实业集团有限公司是浙江泰普森（控股）集团旗下支柱产业公司，始创于1991年。公司产品丰富，有渔具、户外家具、帐篷、花园家具、包袋、打猎等七大系列，品种上万种，产品畅销欧美及亚洲等60多个国家和地区，出口额位列行业前茅，拥有中国合格评定认可委员会认可的国家级检测中心，是中国颇具规模的户外用品生产基地。

Zhejiang Tapson Industrial Group Co., Ltd. is a core industry company under the Zhejiang Tapson（Holdings）Group, founded in 1991. The company offers a wide range of products, including fishing gear, outdoor furniture, tents, garden furniture, bags, hunting equipment, and more, organized into seven major series with thousands of varieties. Its products are popular and widely sold in more than 60 countries and regions, including Europe, the Americas, and Asia. The company's export volume ranks among the top in the industry. It also possesses a National-level Testing Center accredited by the China National Accreditation Service for Conformity Assessment（CNAS）and serves as a significant production base for outdoor products in China.

25. 重庆长安汽车股份有限公司 Chongqing Changan Automobile Co.,Ltd.

重庆长安汽车股份有限公司，简称长安汽车。在重庆、北京、意大利都灵、英国伯明翰等地建立"六国十地"各有侧重的全球协同研发格局。2014年，长安系中国品牌汽车产销累计突破1000万辆，成为第一家跨入"千万俱乐部"的中国品牌。2017年，长安汽车发起第三次创业：创新创业计划，以打造世界一流汽车企业为目标，向智能出行科技公司转型。

Chongqing Changan Automobile Co., Ltd., commonly known as Changan Automobile, is a major Chinese automotive manufacturer. Changan has established a global collaborative research and development framework with a presence in locations such as Chongqing, Beijing, Turin, Italy, and Birmingham, the United Kingdom, known as the "Six Countries and Ten Locations," each with its own specific focus.In 2014, Changan Automobile, as a Chinese brand, achieved cumulative production and sales of over 10 million vehicles, becoming the first Chinese brand to enter the "Ten Million Club." In 2017, Changan Automobile initiated its third venture, the Innovation and Entrepreneurship Program, with the goal of transforming into a world-class automotive enterprise and transitioning into an intelligent mobility technology company.

26. 迈瑞医疗国际股份有限公司 Mindray Medical International Inc.

迈瑞医疗国际股份有限公司是中国领先的高科技医疗设备研发制造厂商，同时也是全球医疗设备的创新领导者之一。自1991年成立以来，迈瑞公司始终致力于临床医疗设备的研发和制造，产品涵盖生命信息与支持、临床检验及试剂、数字超声、放射影像四大领域，将性能与价格完美平衡的医疗电子产品带到世界每一角落。

Mindray Medical is a leading high-tech medical equipment research and manufacturing company in China and one of the innovators in the global medical equipment industry. Since its establishment in 1991, Mindray has been dedicated to the research and manufacturing of clinical medical equipment. Its product portfolio covers four major areas: life information and support, clinical testing and reagents, digital ultrasound, and radiographic imaging. Mindray brings medical electronic products that perfectly balance performance and price to every corner of the world.

27. 吉利集团 GEELY Group

吉利，始建于1986年，1997年进入汽车行业，一直专注实业、技术创新和人才培养，坚定不移地推动企业转型升级和可持续发展。集团以汽车产业电动化和智能化转型为核心，在新能源科技、共享出行、车联网、智能驾驶、车载芯片等前沿技术领域，打造科技护城河，做强科技生态圈。

GEELY, founded in 1986, entered the automotive industry in 1997. The company has always been focused on industry, technological innovation, and talent development, steadfastly promoting corporate transformation, upgrading, and sustainable development. With electric and intelligent transformation of the automotive industry as its core, the group is building a technological moat in cutting-edge technology fields such as new energy technology, shared mobility, vehicle connectivity, intelligent driving, and onboard chips, aiming to strengthen its technological ecosystem.

28. 蔚来汽车 NIO Automobile

蔚来是一家全球化的智能电动汽车公司，于2014年11月25日正式成立。蔚来致力于通过提供高性能的智能电动汽车与极致用户体验，为用户创造愉悦的生活方式。蔚来旗下在售车型包括智能电动旗舰 SUV 全新 ES8、智能电动旗舰轿跑 SUV EC7、智能电动中大型 SUV ES7、智能电动旗舰轿车 ET7、高端智能电动全能 SUV 全新 ES6、智能电动轿跑 ET5，以及智能电动旅行车 ET5T。

NIO is a global smart electric vehicle company that was officially founded on November 25, 2014. NIO is committed to creating a joyful lifestyle for its users by providing high-performance smart electric vehicles and an exceptional user experience. NIO's current lineup of vehicles includes the intelligent electric flagship SUV, the all-new ES8, the intelligent electric flagship coupe SUV EC7, the intelligent electric mid-size SUV ES7, the intelligent electric flagship sedan ET7, the high-end intelligent electric versatile SUV, the all-new ES6, the intelligent electric coupe ET5, and the intelligent electric travel vehicle ET5T.

29. 腾讯控股有限公司 Tecent Holdings Co., Ltd.

腾讯是一家世界领先的互联网科技公司，用创新的产品和服务提升全球各地人们的生活品质。成立于1998年，总部位于中国深圳。公司一直秉承科技向善的宗旨。我们的通信和社交服务连接全球逾10亿人，帮助他们与亲友联系，畅享便捷的出行、支付和娱乐生活。腾讯发行多款风靡全球的电子游戏及其他优质数字内容，为全球用户带来丰富的互动娱乐体验；还提供云计算、广告、金融科技等一系列企业服务，支持合作伙伴实现数字化转型，促进业务发展。

Tencent is a world-leading internet and technology company that develops innovative products and services to improve the quality of life of people around the world.Founded in 1998 with its headquarters in Shenzhen, China, Tencent's guiding principle is to use technology for good. Our communication and social services connect more than one billion people around the world, helping them to keep in touch with friends and family, access transportation, pay for daily necessities, and even be entertained. Tencent also publishes some of the world's most popular video games and other high-quality digital content, enriching interactive entertainment experiences for people around the globe. Tencent also offers a range of services such as cloud computing, advertising, FinTech, and other enterprise services to support our clients' digital transformation and business growth.

30. 华为技术有限公司 UCD 中心 UCD Center, Huawei Technologies Co., Ltd.

华为创立于1987年，是全球领先的 ICT（信息与通信）基础设施和智能终端提供商。我们的20.7万员工遍及170多个国家和地区，为全球30多亿人口提供服务。我们致力于把数字世界带入每个人、每个家庭、每个组织，构建万物互联的智能世界。

HUAWEI, founded in 1987, is a global leader in ICT（Information and Communications Technology）infrastructure and intelligent terminal solutions. With a workforce of over 207,000 employees spread across more than 170 countries and regions, we serve over 3 billion people worldwide. Our mission is to bring the digital world to every individual, family, and organization, and to build an intelligent world where everything is interconnected.

31. 霍尼韦尔 Honeywell

霍尼韦尔是一家《财富》全球500强的高科技企业，为全球提供行业定制的航空产品和服务、楼宇和工业控制技术、以及特性材料，致力于将飞机、汽车、楼宇、工厂、供应链和工人等万物互联，使世界实现更为智能、安全和可持续的长远发展。霍尼韦尔全球总部位于美国北卡罗来纳州夏洛特市，但是一直以来，霍尼韦尔秉持着深耕中国谋求长期发展的理念，贯彻"东方服务东方"和"东方服务世界"的战略，以本土创新推动增长。目前，霍尼韦尔所有业务集团均已落户中国，上海是霍尼韦尔亚太区总部。

Honeywell is a Fortune Global 500 high-tech company that provides industry-customized aviation products and services, building and industrial control technologies, as well as advanced materials worldwide. It is dedicated to connecting everything from planes, cars, buildings, factories, supply chains, and workers to make the world smarter, safer, and more sustainable in the long run. Honeywell's global headquarters is located in Charlotte, North Carolina, USA. However, the company has always adhered to the concept of deepening its presence in China for long-term development, implementing the strategies of "serving the East with the East" and "serving the world with the East," and driving growth through local innovation. Currently, all business groups of Honeywell have established a presence in China, with Shanghai serving as the Asia-Pacific regional headquarters for Honeywell.

鸣谢
ACKNOWLEDGMENT

在此，特别感谢为中国设计智造大奖工作的各位同仁，以及提供过帮助的志愿者们，感谢各位的大力支持。

We would like to sincerely thank all ourcolleagues who have worked for DIA, as well as the volunteers who have contributed. Thank you for supporting us.

机构

浙江省人民政府办公厅、浙江省经济和信息化厅、浙江省财政厅、浙江广播电视集团、浙江省科技厅、浙江省人力资源和社会保障厅、浙江省发展和改革委员会、绍兴市人民政府、杭州市西湖区人民政府、杭州市余杭区人民政府、绍兴市柯桥区人民政府、绍兴市柯桥区科学技术局、余杭梦栖小镇

国际工作站

意大利米兰理工大学 POLI.design 设计学院、欧中设计促进会、奥地利设计协会、韩国设计协会联合会、国际设计研究学会、内罗毕设计周、Blank Group、墨西哥州立自治大学、Core77、Yanko Design、Aspire Ventures Ltd.

协会

世界设计组织、国际设计联合会、中国工业设计协会、中国家用电器协会、中国用户体验行业协会、中国服务型制造联盟、上海设计之都促进中心、浙江省工业设计协会、广东省工业设计协会、福建省工业设计协会、浙江省物联网协会、杭州市工业设计协会、武汉市工业设计协会、泉州工业设计协会、合肥工业设计协会

院校

普瑞特艺术学院、罗切斯特理工学院、新学院大学、加利福尼亚大学伯克利分校、美国南加州大学、英国皇家艺术学院、中央圣马丁大学、伦敦帝国理工学院、伦敦艺术大学、英国拉夫堡大学、伦敦布鲁内尔大学、日本武藏野美术大学、日本千叶大学、德国卡尔斯鲁厄国立设计学院、荷兰埃因霍温科技大学、荷兰代尔夫特理工大学、维也纳应用艺术大学、拉脱维亚艺术学院、韩国国民大学、韩国汉阳大学、韩国弘益大学、韩国蔚山大学、印度哈里亚纳邦国家设计学院、印度珀尔时尚学院

中国美术学院、中央美术学院、清华大学美术学院、浙江大学、同济大学、南京艺术学院、四川美术学院、广州美术学院、实践大学、香港科技大学、香港理工大学、湖南大学、吉林大学、苏州大学、北京科技大学、北京服装学院、江南大学、浙江工业大学、浙江理工大学、浙江工商大学、广东工业大学、陕西科技大学、湖北工业大学、武汉理工大学、郑州轻工业大学、燕山大学、湖南工业大学、内蒙古科技大学、福建工程学院、华北理工大学、吉林艺术学院、江西师范大学、山东工艺美术学院、电子科技大学中山学院

媒体

《装饰》杂志、《设计》杂志、网易设计、《艺术与设计》杂志、新浪家居、设计赛、设计与制造

新华网、中新网、人民网、央广网、光明网、中国日报网、中国经济网、中国青年网、中国财经时报网、国际在线、环球网、北青网、中华网、澎湃新闻、浙江新闻、浙江在线、浙江发布、天目新闻、新蓝网、杭州网、网易新闻、搜狐网、新浪网、凤凰网、界面新闻、Zaker 新闻、世界浙商网、腾讯家居、太平洋家居网、DONEWS、TECHWEB、中文科技资讯、科技讯、科技网、飞象网、赛迪网、矮凳网

新华社、人民日报、解放日报、浙江日报、新京报、中国青年报、中国工业报、消费日报、中国文化报、都市快报、钱江晚报、杭州日报、青年时报、经济日报、中国美术报、美术报、上海日报、深圳商报、深圳特区报、深圳晶报、深圳晚报、每日商报、文汇报、科技金融时报、新民周刊、浙商杂志、文化产业月刊、时尚周末

CCTV-4中文国际频道、中国蓝 TV、浙江电视台钱江频道、浙江电视台公共新闻频道、浙江电视台教育科技频道、杭州电视台综合频道、杭州电视台西湖明珠频道、中央人民广播电台、FM95浙江经济广播、FM93 交通之声、FM88 浙江之声、FM89 杭州之声

彭博社、美联社、路透社、法新社、俄塔社、雅虎财经、NBC 全国广播公司、The Times、道琼斯市场观察、加拿大《财富》、澳洲《商业日报》、今日美国、晨星网、BuzzFeed、雅虎新闻、世界新闻网、美国科技日报、欧洲邮政公报、布法罗新闻报、美国先锋晚报、加拿大数字期刊、英国城市新闻报、亚利桑那共和报、每日先驱日报、美国哥伦比亚广播、美国福克斯电视台、亚洲第一站、CW 电视台、Core77、Yanko Design

Institutions:
The People's Government of Zhejiang Province, Economy and Information Technology Department of Zhejiang, Zhejiang Provincial Department of Finance, Zhejiang Radio & TV Group, Science Technology Department of Zhejiang Province, Zhejiang Province Human Resources and Social Security Department, Zhejiang Provincial Development and Reform Commission, Shaoxing Municipal Government, the People's Government of Hangzhou Xihu District,the People's Government of Hangzhou Yuhang District, the People's Government of Shaoxing Keqiao District, Shaoxing Keqiao Science and Technology Bureau, Hangzhou Liangzhu New City Mengqi Town

Internationl Workstations:
POLI. design Società consortile a responsabilità limitata, Stichting voor Design Promotion in Europa en China, Designaustria, Korean Federation of Design Associations, International Forum for Design Research, Nairobi Design Week, Blank Group, Autonomous University of the State of Mexico, Core77, Yanko Design, Aspire Ventures Ltd.

Associations:
World Design Organization, International Council of Design, China Industrial Design Association, China Household Electrical Appliances Association, User Experience Professional Association China, China Service-oriented Manufacturing Alliance, Shanghai Promotion Center for City of Design, Zhejiang Industrial Design Association, Guangdong Industrial Design Association, Fujian Industrial Design Association, Zhejiang Association for IOT Industry, Hangzhou Industrial Design Association, Wuhan Industrial Design Association, Quanzhou Industrial Design Association, Hefei Industrial Design Association

Universities:
Pratt Institute, Rochester Institute of Technology, The New School, University of California, Berkeley, University of Southern California, Royal College of Art, Central Saint Martins, Imperial College London, University of the Arts London, Loughborough University, Brunel University London, Politecnico di Milano, Musashino Art University, Chiba University, Karlsruhe University of Art and Design, Eindhoven University of Technology, Delft University of Technology, University of Applied Arts Vienna, Art Academy of Latvia, Kookmin University, Hanyang University, Hongik University, University of Ulsan, National Institute of Design, Pearl Academy of Fashion

China Academy of Art, Central Academy of Fine Arts, Academy of Arts & Design,Tsinghua University, Zhejiang University, Tongji University, Nanjing University of the Arts, Sichuan Fine Arts Institute, Guangzhou Academy of Fine Arts, Shih Chien University, The Hong Kong University of Science and Technology, The Hong Kong Polytechnic University, Hunan University, Jilin University, Soochow University, University of Science and Technology Beijing, Beijing Institute of Fashion Technology, Jiangnan University, Zhejiang University of Technology, Zhejiang Sci-Tech University, Zhejiang Gongshang University, Guangdong University of Technology, Shaanxi University of Science & Technology, Hubei University of Technology, Wuhan University of Technology, Zhengzhou University of Light Industry, YanShan University, Hunan University of Technology, Inner Mongolia University of Science and Technology, Fujian University of Technology, North China University of Science and Technology, Jilin University of Arts, Jiangxi Normal University, Shandong University of Art & Design, and University of Electronic and Technology of China, Zhongshan Institute

Media:
ZHUANGSHI, Design, NetEase Home, Art and Design, Sina Home, Design Competition, Design & Manufacture

Xinhua Net, China News, People's Daily Online, CNR News, Guangming Online, China Daily, China Economic Net, Youth.cn, 3News.cn, CRI Online, Huanqiu.com, YNET.com, China.com, The Paper, Zhejiang News, Zhejiang Online, Zhejiang Release, Tianmu News, CZTV.com, Hangzhou.com.cn, NetEaseNews, Sohu.com, Sina.com, IFENG.com, JIEMIAN.com, ZakerNews, WZS.org.cn, J360.com, PCHOUSE, DONNEWS, TechWeb, Citnews.com.cn, Kejixun.com, Kejiwang.com.cn, Cctime.com, CCIDNET.com, IDEN.cn

Xinhua News Agency, People's Daily, Jiefang Daily, Zhejiang Daily, The Beijing News, China Youth Daily, China Industry News, Consumption Daily, China Culture Daily, Dushi Kuaibao, Qianjiang Evening News, Hangzhou Daily, Youth Times, Economic Daily, Art News of China, China Art Weekly, Shanghai Daily, Shenzhen Economic Daily, Shenzhen Special Zone Daily, Daily Sunshine, Shenzhen Evening News, Daily Business, Wenhui Daily, Sci-Tec & Finance Times, Xinmin Weekly, Zhe Shang Magazine, Zhejiang Culture Industry Guide, Style Weekend

CCTV-4, China-blue TV, Qianjiang channel of Zhejiang Radio and Television Group, Public News channel of Zhejiang Radio and Television Group, Education and Technology channel of Zhejiang Radio and Television Group, Multi-format channel of Zhejiang Radio and Television Group, General Channel of Hangzhou TV Station, West Lake Pearl channel of Hangzhou TV Station, China National Radio, FM 95 Zhejiang Economic Broadcasting, FM93 The sound of traffic, FM88 The Voice of Zhejiang, FM89 The Voice of Hangzhou

Bloomberg, Associated Press, Reuters, Agence France-Presse, TASS, Yahoo Finance, NBC National Broadcasting Company, The Times, MarketWatch, Canadian Business, The Australian Financial Review, USA Today, Morningstar, BuzzFeed, Yahoo News, World Journal, U.S. News & World Report, European Postal Bulletin, The Buffalo News, The Evening Leader, Digital Journal, City A.M. (London), The Arizona Republic, Daily Herald, CBS News, Fox Broadcasting Company, AsiaOne, The CW Television Network, Core77, Yanko Design

专家（按姓氏首字母排序）

日本：
Aoki Shiro, Shigenori Asakura, Hideichi Misono, Kota Nezu, Eisuke Tachikawa, Makoto Watanabe, Nobuyoshi Yamazaki,Takashi Yamada

韩国：
Cha Kang-Heui, Chin Byung-wook, Kun-pyo Lee, Kim Hyunsun

法国：
Nicolas Cinguino, Fabien Grégoire, Patrick Lucien Lapoule, Adrien NAZEZ, Jean Christophe NAOUR, Stéphane Thirouin

美国：
Deb Aldrich, Allan Chochinov, Laura Des Enfants, Ravindra S.Goonetilleke, Lorraine Justice, Long Jiao, Tong Jin Kim, Beau Oyler, Brian Roderman, Srini R. Srinivasan, Kostas Terzidis

英国：
Amanda Broderick, James Hope, Stephen Green, Russ Shaw, Mark Hedley, Adrian Jankowiak,John Mathers, James Vanderpant

德国：
Moritz Ludwig

奥地利：
Martin Foessleitner, Severin Filek, Florian Halm, Anna Maislinger, Mathias Schasché

荷兰：
Loe Fejis, Marina Toeters

马达加斯加：
Fenoson Zafifimahova

比利时：
Aloke B. Nandi

意大利：
Michele Brunello, Cristian Confalonieri, Luca Fois, Francesco Zurlo

以色列：
David William Grossman

芬兰：
Tapani Hyvönen

印度：
Balkrishna Mahajan, Radhika Seth, Sarang Sheth, Pradyumna Vyas

土耳其：
Eray Sertac Ersayin, Ilgaz Kuruyazici, Fatih Mintas

墨西哥：
Consuelo Espinosa, Francisco Santín, Martha Zarza

肯尼亚：
Mugendi K. M' Rithaa

中国：
毕学锋、常冰、蔡军、陈妍、曹杰、陈汗青、陈江、段胜峰、段卫斌、方晓风、方海、郭春方、胡晓、韩绪、胡志鹏、胡飞、洪华、何人可、韩挺、胡军、季铁、蒋春晖、贾京生、贾荣林、官政能、郭介诚、李杰、卢刚、刘征、兰翠芹、林敬亭、雷海波、鲁晓波、李思卫、刘小康、刘宁、卢纯福、娄永琪、李超德、李加林、覃京燕、邵景峰、孙守迁、孙凌云、宋协伟、宗福季、童慧明、陶小年、魏洁、吴卓浩、王路平、徐冰、杨学太、杨光、余隋怀、朱旭光、支锦亦、周红石、朱仁民、张凌浩、赵超、张明、踪雪梅、章俊杰、赵业、周立钢、张展、赵健

支持单位：
浙江省人民政府

主办单位：
中国美术学院

协办单位：
中国工业设计协会
教育部高等学校工业设计专业教学指导分委员会

承办单位：
中国设计智造大奖组委会

运营单位：
浙江现代智造促进中心

Experts (Alphabetical Order)

Japan:
Aoki Shiro, Shigenori Asakura, Hideichi Misono, Kota Nezu, Eisuke Tachikawa, Makoto Watanabe, Nobuyoshi Yamazaki,Takashi Yamada

South Korea:
Cha Kang-Heui, Chin Byung-wook, Kun-pyo Lee, Kim Hyunsun

France:
Nicolas Cinguino, Fabien Grégoire, Patrick Lucien Lapoule, Adrien NAZEZ, Jean Christophe NAOUR, Stéphane Thirouin

USA:
Deb Aldrich, Allan Chochinov, Laura Des Enfants, Ravindra S. Goonetilleke, Lorraine Justice, Long Jiao, Tong Jin Kim, Beau Oyler, Brian Roderman, Srini R. Srinivasan, Kostas Terzidis

UK:
Amanda Broderick, James Hope, Stephen Green, Russ Shaw, Mark Hedley, Adrian Jankowiak,John Mathers, James Vanderpant

Germany:
Moritz Ludwig

Austria:
Martin Foessleitner, Severin Filek, Florian Halm, Anna Maislinger, Mathias Schasché

Netherlands:
Loe Feijs, Marina Toeters

Madagascar:
Fenoson Zafifimahova

Belguim:
Aloke B. Nandi

Italy:
Michele Brunello, Cristian Confalonieri, Luca Fois, Francesco Zurlo

Israel:
David William Grossman

Finland:
Tapani Hyvönen

India:
Balkrishna Mahajan, Radhika Seth, Sarang Sheth, Pradyumna Vyas

Turkey:
Eray Sertac Ersayin, Ilgaz Kuruyazici, Fatih Mintas

Mexico:
Consuelo Espinosa, Francisco Santín, Martha Zarza

Kenya:
Mugendi K. M' Rithaa

China:
Bi Xuefeng, Chang Bing, Cai Jun, Chen Yan, Cao Jie, Chen Hanqing, Chen Jiang , Duan Shengfeng, Duan Weibin, Fang Xiaofeng , Fang Hai, Guo Chunfang, Hu Xiao, Han Xu, Hu Zhipeng, Hu Fei, Hong Hua, He Renke, Han Ting, Hu Jun, Ji Tie, Jiang Chunhui, Jia Jingsheng, Jia Ronglin, Guan Zhengneng, Guo Jiecheng, Li Jie, Lu Gang, Liu Zheng, Lan Cuiqin, Lin Jingting, Lei Haibo, Lu Xiaobo, Lee Sherman, Lau Freeman, Liu Ning, Lu Chunfu, Lou Yongqi, Li Chaode, Li Jialin, Qin Jingyan, Shao Jingfeng, Sun Shouqian, Sun Lingyun, Song Xiewei, TSUNG Fugee, Tong Huiming, Tao Xiaonian, Wei Jie, Wu Zhuohao, Wang Luping, Xu Bing, Yang Xuetai, Yang Guang, Yu Suihuai, Zhu Xuguang, Zhi Jinyi, Zhou Hongshi, Zhu Renmin, Zhang Linghao, Zhao Chao , Zhang Ming, Zong Xuemei, Zhang Junjie, Zhao Ye, Zhou Ligang, Zhang Zhan , Zhao Jian

Supporting Unit:
The People's Government of Zhejiang Province

Host:
China Academy of Art

Co-organizer:
China Industrial Design Association / Steering Sub-committee on the Teaching of Industrial Design in Higher Educational Institutions under the Ministry of Education

Organizer:
Design Intelligence Award Committee

Operating Unit:
Zhejiang Modern Intelligence and Manufacturing Promotion Center

作品索引
ENTRY INDEX

U: 浙江本来家居科技有限公司 / 杭州本来工业设计有限公司
belaDESIGN wood / belaDESIGN

P107 趣味八段锦叠叠乐
BUILDING BLOCK TOY OF THE EIGHT BROCADES
A: 陆明朗 魏婷
Lu Minglang , Wei Ting
U: 宁波美乐雅荧光科技股份有限公司 / 厦门市拙雅科技有限公司
Ningbo Merryart Glow-tech Co.,Ltd. / Xia Men Joyatech Co., Ltd.

P108 SECOND SKINS
A: Malou Beemer, Christian Dils
U: Malou Beemer, Atelier Mlou

P108 IICONIC 鞋
IICONIC FOOTWEAR
A: Iris Camps
U: Eindhoven University of Technology

P109 JUCK
A: Yongil Lee, Jaemyung Seo, Bonghee Kim
U:Artworks Group

P109 FLIPPACK 防水防盗折叠背包
FLIPPACK
A: 黄辉古 赵壁 曾繁渠
Huang Huigu, Zhao Bi, Zeng Fanqu
U: 广州市阔云科技有限公司
Guangzhou Korin Technology Co., Ltd.

P110 净酌
JO-CHU
A: Eisuke Tachikawa, Ryota Mizusako, Jin Nagao, Daichi Komatsu, Naoki Hijikata
U: NAORAI Co., Ltd. NOSIGNER

P110 相见欢温酒器
XIANG JIAN HUAN WINE WARMER
A: 郑润桦 黄俊杰 兰鼎平
Zheng Runhua , Huang Junjie , Lan Dingping
U: 南山先生（厦门）文化创意有限公司
Mr. Nan Shan (Xiamen) Cultural Creativity Co., Ltd.

P111 传家壶 —— 迷你
HEIRLOOM THERMOS — MINI
A: 季坤荣 夏飞剑 孔洪强 刘安民 吴冬
Ji Kunrong, Xia Feijian , Kong Hongqiang , Liu Anmin , Wu Dong
U: 深圳市一原科技有限公司
ShenZhen Yiyuan Technology Co., Ltd

P111 雅琮提梁壶
YA CONG HANDLE POT
A: 徐乐 包力源 辜弯婉 陈欣 沈堃昊
Xu Le, Bao Liyuan, Gu Wanwan, Chen Xin, Shen Kunhao
U: 杭州大巧创意设计有限公司 / 浙江工业大学之江学院
Hangzhou Great Wisdom Creative Design Co., Ltd. / Zhijiang College of Zhe jiang University of Technology

P112 SOFTLINE
A: Ahmet Osman PEKER
U: Kar porselen

P112 消块 模块家具
MAP STOOL
A: PD A, PD B, PD C
U: GYRO, PARTISAN 404

P113 珐琅铁铸饭煲
ENAMELLED IRON CAST RICE COOKER
A: 张必锋 杨扬 杨能鹏 卢传德 罗玮瑜
Zhang Bifeng, Yang Yang, Yang Nengpeng, Lu Chuande, Luo Weiyu
U: 广东顺德米壳工业设计有限公司
MIKO Industrial Design Co.,Ltd.

P113 一席
PRIVATE TEA SPACE
A: 磨炼 王兆巧 陈雯
Mo Lian, Wang Zhaoqiao, Chen Wen
U: 深圳市聿上家具设计有限公司 / 广州美术学院
YOSON Design Co., Ltd. / Guangzhou Academy of Fine Arts (GAFA)

P114 PLATO COLLECTION
A: Kerem Aris, Merve Parnas, Defne Koz
U: Uniqka

P114 KONI
A: Kerem Aris, Merve Parnas, Romy Kühne
U: Uniqka

P115 戏出东方
SHADOW PLAY IN CHINA
A: 张书雁 王猛涛
Zhang Shuyan, Wang Mengtao
U: 浙江自然造物文化创意有限公司
Made in nature

P115 搭扣
BUCKLE
A: 姜军 金炜楠
Jiang Jun, Jin Weinan
U: 上海瑞鹊投资有限公司 / 杭州乐造工业设计有限公司
Shanghai RUIQUE Investment Co., Ltd. / Hangzhou LeDesign Co., Ltd.

P116 玲珑架
LINGLONG SHELF
A: 周安彬
Zhou Anbin
U: 深圳得闲设计有限公司
Shenzhen DeXian Design Co., Ltd.

P116 M+
A: 薛平安 李超阳
Xue Pingan，Li Chaoyang
U: 致欧家居科技股份有限公司
Ziel Home Furnishing Technology Co., Ltd.

P117 PER 系列
PER SERIES
A: Miyu Ikeda , Takuto Kurashima
U: Hirata Chair Manufacture Co., Ltd.

P117 MANA
A: Ozan Tığlıoğlu , Wakako Esra Aras , Furkan Öz
U: GOA Design Factory

P118 熊抱
BEAR HUG
A: 缪景怡 林丹慧 尹雪
Miao Jingyi, Lin Danhui, Yin Xue

P118 INFINITE GROWTH
A: Jun Wang, Xia Hua, Lingxiao Zeng, Yining Xu, Mingdong Mao
U: Junwang Studio

P119 这里有猫腻
THERE IS SOMETHING GOING ON
A: 吴曲
Wu Qu

P119 铣型椅
CNC
A: Yrjo Kukkapuro
U: 上海阿旺特家具有限公司
Shanghai Avarte Furniture Co.,Ltd.

P120 BAN
A: Yong-il Lee, Jaemyeong Seo, Bonghee Kim
U: Minjakso

P120 RELICS
A: Georgia von le Fort

P121 "变"几
CHANGEABLE TABLE
A: 李雅云 孙雪松 兰钊 王凤茹 林永贤
Li Yayun, Sun Xuesong, Lan Zhao, Wang Fengru, Lin Yongxian
U: 中德（泉州）工业设计研究院有限公司
Sino German (Quanzhou) Industrial Design and Research Institute Co., Ltd.

P121 F=MG
A: Jakob Glasner

P122 心冥想座具便携系列·小忍
LITTLE PATIENCE [MEDITATION SEAT[PORTABLE]]
A: 高凤麟，刘沛桐
Gao Fenglin, Liu Peitong
U: 心冥想健康科技有限公司
Shine Meditation Health Technology （Hangzhou）Co., Ltd.

P122 笑椅
SMILE — CHAIR
A: 徐乐 包力源 瞿伟民 林幸民 辜弯婉
Xu Le, Bao Liyuan, Zhai Weimin, Lin Xingmin, Gu Wanwan
U: 杭州大巧创意设计有限公司 / 浙江工业大学之江学院
Hangzhou DAQIAO Home Design Studio/Zhijiang College of Zhejiang University of Technology

P123 12° Lamp
A: Kun Geng
U: FOM STUDIO SRLS

P123 FITNESS PARCOURS
A: Sigi Ramoser, Nico Pritzl

P124 随"变"换
CHANGEABLE LENS ADJUSTABLE PUPIL DISTANCE GLASSES
A: 左叶 张森
Zuo Ye, Zhang Sen

U: 温州市森一眼镜设计有限公司
Senee Eyewear Design Studio

P124 再生混凝土海洋消波块微生态设计
MICRO ECOLOGICAL DESIGN OF RECYCLED CONCRETE OCEAN WAVE ABSORBING BLOCK
A: 龙香华 刘佳豪 谢仁科
Long Xianghua, Liu Jiahao, Xie Renke

P125 厕田
FIELD SANITATION UNIT
A: 杨窝 陈向阳
Yang Jun, Chen Xiangyang
U: 杭州玩家家居设计有限公司
Hangzhou Spieler Design Pty. Ltd.

P125 BEACH ENVELOPED IN MIST
A: Reiko Kitora, Atsuhito Kitora

P126 THINKPAD PLASTIC FREE PACKAGING
A: Lenovo Design Innovation Team
U: Lenovo

P126 SUSTAINABLE SHOPPING BAG
A: Lim Sungmook

P127 常青
EVERGREEN
A: 杜聪 张湘钰 崔喆 米男男 张书炜
Du Cong, Zhang Xiangyu, Cui Zhe, Mi Nannan, Zhang Shuwei
U: 合肥联宝信息技术有限公司
Hefei LCFC Information Technology Co., Ltd.

P127 KENYAN UPCYCLED UNIFORMS
A: Alex Musembi, Elmar Stroomer, Kirsty Zeller, Khalid Awale
U: Africa Collect Textiles

P128 三江源雪域牦牛绒围巾公益礼品
SANJIANGYUAN CHARITY GIFTS
A: 黄莎莉 张安吉 曾令波
Huang Shali, Zhang Anji, Zeng Lingbo
U: 腾讯科技（深圳）有限公司 / 深圳慢物质文化创意有限公司
Tencent Technology (Shenzhen) Co., Ltd. / Shenzhen Slow Material Culture Creative Co., Ltd.

P128 循环直运快递箱 —— 为快递包装"绿色化"发展而设计
INFINITE LOOP BOX — "GREEN" FOR EXPRESS PACKAGING
A: 李甫印 江佳婧 陈家超
Li Fuyin, Jiang Jiajing, Chen Jiachao
U: 中荣印刷集团股份有限公司
Zrp Printing Group Co.,Ltd.

P129 RE-CARDBOARD
A: Reiko Kitora, Atsuhito Kitora

P129 "莲花岛号"海洋鱼类培育放生船
"LOTUS ISLAND" MARINE FISH CULTIVATION AND RELEASE SHIP
A: 朱仁民
Zhu Renmin
U: 杭州潘天寿环境艺术设计有限公司
Hangzhou Pantianshou Environmental Art Design Co., Ltd.

Hugo Richardson
U: The Tyre Collective

P198 儿童无针注射器
NEEDLE-FREE INJECTOR FOR KIDS
A: 陈苏宁
Chen Suning
U: 北京快舒尔医疗技术有限公司
Beijing QS Medical Technology Co.,Ltd.

P199 新华医疗智慧化 CSSD 整体解决方案
THE OVERALL PLAN OF SHINVA
INTELLIGENT DISINFECTION SUPPLY
CENTER
A: 王泽坤 韩建康 李现刚 朱书建 任义凯
Wang Zekun, Han Jiankang, Li
Xiangang, Zhu Shujian, Ren Yikai
U: 新华医疗器械股份有限公司
SHINVA

P199 小红花微针疫苗
A LITTLE RED FLOWER
A: 缪景怡 邹洈
Miao Jingyi, Zou Hu

P200 幻肢
V-HANDER
A: 王翌诚
Wang Yicheng
U: 浙江工业大学设计与建筑学院
School of Design and Architecture,
Zhejiang University of Technology

P200 DNR AUTO ANGLE
A: Ahmet brahim POLAT, Aamish
UMAR , Kanber SEDEF
U: DENER MAKNA SAN. VE TC. LTD.ŞT.

P201 C 型移动式 X 光射线机
C-ARM X-RAY
A: 张培培 滕轩 陈泽 卢恒
Zhang Peipei, Teng Xuan, Chen Ze, Lu
Heng.
U: 南京佗道医疗科技有限公司
Nanjing Tuodao Medical Technology Co.,
Ltd.

P201 V3 呼吸机
V3 VENTILATOR
A: 钟琳琳 匡思能 周妙雨
Zhong Linlin, Kuang Sineng, Zhou
Miaoyu
U: 深圳市科曼医疗设备有限公司
Shenzhen Comen Medical Instruments
Co., Ltd.

P202 鱼跃电子体温计 YT-3
YUWELL ELECTRONIC
THERMOMETER YT-3
A: 施逸琪 杨天照 华昊 邱博 高昱明
Shi Yiqi, Yang Tianzhao, Hua Hao, Qiu
Bo, Gao Yuming
U: 江苏鱼跃医疗设备股份有限公司
JIANGSU YUWELL MEDICAL
EQUIPMENT & SUPPLY Co., Ltd.

P202 鱼跃快速检测系列卡壳
YUWELL RAPID SELF-TEST KITS
A: 郑怡珺 华昊 施逸琪 赵扬 杨天照
Zheng Yijun, Hua Hao, Shi Yiqi, Zhao
Yang, Yang Tianzhao
U: 江苏鱼跃医疗设备股份有限公司

JIANGSU YUWELL MEDICAL
EQUIPMENT & SUPPLY Co., Ltd.

P203 鱼跃臂式一体血压计 630CR
YUWELL UPPER ARM BLOOD
PRESSURE MONITOR 630CR
A: 杨天照 施逸琪 高昱明 卞程莹 华昊
Yang Tianzhao, Shi Yiqi, Gao Yuming,
Bian Chengying, Hua Hao
U: 江苏鱼跃医疗设备股份有限公司
JIANGSU YUWELL MEDICAL
EQUIPMENT & SUPPLY Co., Ltd.

P203 鱼跃安耐糖 CT3 持续葡萄糖监测系统
YUWELL CT3 CONTINUOUS GLUCOSE
MONITORING SYSTEM
A: 孙博珍 华昊 史小雅 施逸琪 杨天照
Sun Bozhen, Hua Hao, Shi Xiaoya, Shi
Yiqi, Yang Tianzhao
U: 江苏鱼跃医疗设备股份有限公司
JIANGSU YUWELL MEDICAL
EQUIPMENT & SUPPLY Co., Ltd.

P204 华诺康 4K 内窥镜系统
THE HEALNOC 4K ENDOSCOPY
SYSTEM
A: 魏亚军 陈力 卢强博 王胜浩 靳华玲
Wei Yajun, Chen Li, Lu Qiangbo, Wang
Shenghao, Jin Hualing
U: 浙江华诺康科技有限公司 / 浙江大华
技术股份有限公司
Zhejiang Healnoc Technology Co.,Ltd. /
Dahua Technology Co., Ltd.

P204 悠然 210 下肢坐卧式康复训练器
URA210 LOWER LIMB REHABILITATION
TRAINER
A: 颜海 王天 赵晴宇 李聪 叶广兴
Yan Hai, Wang Tian, Zhao Qingyu, Li
Cong, Ye Guangxing
U: 杭州程天科技发展有限公司
Hangzhou RoboCT Technology
Development Co.,Ltd

P205 智能动力防水小腿假肢
WATERPROOF SMART POWERED
TRANSTIBIAL PROSTHESIS P105
A: 罗嶷 汤晓杭
Luo Yi, Tang Xiaohang
U: 北京工道风行智能技术有限公司 / 深
圳市摩迪赛产品设计管理有限公司
Beijing Speedsmart Technology Co.,
Ltd. / Shenzhen Modesign Management
Co., Ltd.

P205 X66 麻醉工作站
X66 ANESTHESIA SYSTEM
A: 赵东阁 王延庆 陈定亮 刘国新 魏璟涛
Zhao Dongge, Wang Yanqing, Chen
Dingliang, Liu Guoxin, Wei Jingtao
U: 北京思瑞德医疗器械有限公司 / 北京
智加问道科技有限公司
Beijing Siriusmed Medical Equipment
Co., Ltd. / ZCO Design Co., Ltd.

P206 Typdont+
A: Luke Goh Xu Jie, Loh Yue Xuan
Chantel, Hugo Jean Guillaume
U: National University of Singapore
(NUS)

P206 下肢康复机器人
LOWER LIMB REHABILITATION ROBOT
A: 赵智峰 芮达志 唐聪智 陈亚新 李得阳
Zhao Zhifeng, Rui Dazhi, Tang Congzhi,
ChenYaxin, Li Deyang
U: 上海璟和技创机器人有限公司 / 苏州
柒整合设计有限公司
Shanghai Jinghe Technology Robotics
Co., Ltd. / Suzhou CHY Integration
Design Co., Ltd.

P207 ARTEMIS 389
A: Anuja Tripathi

P207 BOB 儿童康复平衡推举仪
BOB CHILDREN'S REHABILITATION
BALANCE PRESS
A: 任紫涵 王龙龙
Ren Zihan, Wang Longlong

P208 智慧医疗 —— 下肢康复穿戴设备
WITMED — LOWER LIMB
REHABILITATION WEARABLE DEVICES
A: 秦桂祥 许佳 胡浩伟
Qin Guixiang, Xu Jia, Hu Haowei

P208 磁共振引导肝癌消融智能手术机器人
MRI GUIDED SURGICAL ROBOT
DESIGN FOR LIVER CANCER
A: 孙博文 袁少玫 李迪嘉 汤毅 王嘉
Sun Bowen, Yuan Shaomei, Li Dijia,
Tang Yi, Wang Jia
U: 北京精准医械科技有限公司 / 北京理
工大学设计与艺术学院
Beijing Precision Medtech Technologies
Co.,Ltd. / School of Design &
Arts,Beijing Institute of Technology

P209 AR 骨科手术导航系统
HOLONAVI LUMBAR PUNCTURE
OUTFIT
A: 高凤麟 刘沛桐 刘洋
Gao Fenglin, Liu Peitong, Liu Yang
U: 上海霖晏医疗科技有限公司 / 心冥想
健康科技 (杭州) 有限公司
Shanghai Linyan Medical Technology
Co., Ltd. / Shine Meditation Health
Technology (Hangzhou) Co., Ltd.

P209 载人 eVTOL 飞行器（ZG-ONE）
MANNED EVTOL AIRCRAFT（ZG-ONE）
A: 贾思源 李宜恒 班剑锋 黄宇欣 陈邵洁
Jia Siyuan, Li Yiheng, Ban Jianfeng,
Huang Yuxin, Chen Shaojie
U: 零重力深圳飞机工业有限公司
Zero-G Shenzhen Aircraft Industry
Co.,Ltd.

P210 NKB5000—— 视控一体键盘
NKB5000 KEYBOARD
A: 刘雅慧 陈力 李玮 杨康 彭皓明
Liu Yahui, Chen Li, Li Wei, Yang Kang,
Peng Haoming
U: 浙江大华技术股份有限公司
Dahua Technology Co., Ltd.

P210 基于并联旁路动力变速的铰接式山地拖
拉机
ARTICULATED MOUNTAIN TRACTOR
A: 周宁 罗钦林 李荣 廖建群 唐刚
Zhou Ning, Luo Qinlin, Li Rong, Liao
Jianqun, Tang Gang
U: 长沙九十八号工业设计有限公司
N98Design

P211 VOGUE HIGHBAY
A: Sushmita Jaiswal, Sumit Singh, Deep
Singh

P211 聋哑人手势智能翻译臂环
INTELLIGENT GESTURE ARMBAND
FOR THE DEAF
A: 滕佳琪 曾芷涵 徐子昂 夏涵飞 厉向东
Teng Jiaqi, Zeng Zhihan, Xu Ziang, Xia
Hanfei, Li Xiangdong
U: 浙江大学
Zhejiang University

P212 三一电动重卡车头
SANY ELECTRIC HEAVY TRUCK HEAD
A: 朱宏 吕振广 刘海军 陈勇 郭海亮
Zhu Hong, Lv Zhenguang, Liu Haijun,
Chen Yong, Guo Hailiang
U: 三一集团有限公司
SANY Group Co.,Ltd.

P212 极 22 三维扫描仪
EXTR22 3D SCANNER
A: 盛明圆 茹方军 李冠楠
Sheng Mingyuan, Ru Fangjun, Li
Guannan
U: 杭州非白三维科技有限公司
FORMBUILDER

P213 VT-30
A: 胡华智 李智奕 马春明 陈娜
Hu Huazhi, Li Zhiyi, Ma Chunming,
Chen Na
U: 亿航智能设备（广州）有限公司
EHang

P213 TO SEE OR NOT TO SEE
A: Tuncay ince
Tuncay ince

P214 HYDROSURV
A: David Hull
U: HydroSurv

P214 方圆号 ——75m 东方美学商务游艇概
念设计
FANG YUAN—75M ORIENTAL
BUSINESS YACHT CONCEPT
A: 曹中淏 王萌 李博 吴亮 孙天为
Cao Zhonghao, Wang Meng, Li Bo, Wu
Liang, Sun Tianwei

P215 未来智能交通巴士
FUTURE INTELLIGENT
TRANSPORTATION BUS
A: 刘卉媛
Liu HuiYuan

P215 飞凡 R7
RISING AUTO R7
A: 邵景峰 邵长山 齐精文 王睿 黄晴辉
Shao Jingfeng, Shao Changshan, Qi
Jingwen, Wang Rui, Huang Qinghui
U: 上汽集团 / 上汽设计中心
SAIC GROUP / SAIC DESIGN CENTER

P216 可持续清洁公共洗车机器人系统
SUSTAINABLE CLEANING PUBLIC CAR
WASH ROBOT SYSTEM
A: 刘志强 林树毫
Liu Zhiqiang, Lin Shuhao
U: 广州美术学院
Guangzhou Academy of Fine Arts

P216 美团 X1 共享电单车
MEITUAN X1 SHARING E-SCOOTER
A: 刘译元 赵猛 章思远 邹慧琳 毛非一
Liu Yiyuan, Zhao Meng, Zhang Siyuan,
Zou Huilin, Mao Feiyi
U: 美团
Meituan

P217 818防疫安检通道
ULTRA ENTRANCE GATE
A: 薛李安 陈力 杨易铭 李示明 黄祝秋
Xue Li'an, Chen Li, Yang Yiming, Li
Shiming, Huang Zhuqiu
U: 浙江华视智检科技股份有限公司 / 浙
江大华技术股份有限公司
Zhejiang Huajian Technology Co.,Ltd. /
Dahua Technology Co., Ltd.

P217 双差速舵轮驱动重载 AGV
DOUBLE DIFFERENTIAL HELM DRIVES
HEAVY DUTY AGV
A: 范育芳 莫光辉 胡徐起 叶丹丹 吕荣平
Fan Yufang, Mo Guanghui, Hu Xuqi, Ye
Dandan, Lv Rongping
U: 汉度（杭州）创意设计发展有限公司
Hando Design

P218 范德兰德 SBD 机场自助值机系统
VANDERLANDE SBD AIRPORT SELF
CHECK-IN SYSTEM
A: 马海豹 朱池 刘昆 顾闻 郑文雄
Ma Haibao, Zhu Chi, Liu Kun, Gu Wen,
Zheng Wenxiong
U: 范德兰德物流自动化系统（上海）
有限公司 / 上海木马工业产品设计有
限公司
Vanderlande Industries Logistics
Automated Systems (Shanghai) Co., Ltd.
/ Shanghai Muma Industrial Products
Design Co., Ltd.

P218 为载人登火任务服务的子母火星车系统
概念设计
MOTHER-CHILD MARS ROVER FOR
MANNED FIRE MISSIONS
A: 李自翔
Li Zixiang

P219 SD-10
A: Hitoshi Igarashi, Manabu Kawahara,
Hideki Kato, Shigeaki Isobe
U: SEIKO EPSON CORPORATION

P219 THE HOMEOWNER'S HYDROPONICS
A: Nahita Zafimahova

数字经济 DIGITAL ECONOMY

P221 支付宝盒 R0
ALIPAY R0
A: 张铭伟 陈志远
Zhang Mingwei, Chen Zhiyuan
U: 支付宝
Alipay

P221 微信刷掌服务
WEPALM
A: 张颖 叶娃 侯锦坤 郭润增 黄家宇
Zhang Ying, Ye Wa, Hou Jinkun, Guo
Runzeng, Huang Jiayu
U: 腾讯科技（深圳）有限公司
Tencent Technology (Shenzhen) Co.,Ltd.

P222 魔派系列智能终端
MAGIC PAD SMART TERMINAL
A: 黄立清 廖鹭蓉 蔡玉敏
Huang Liqing, Liao Lurong, Cai Yumin
U: 厦门立林科技有限公司
XIAMEN LEELEN TECHNOLOGY Co., Ltd.

P222 WATERFALL-NYC
U: D'STRICT KOREA, INC.

P223 吉利银河车机系统
GEELY GALAXY OS
A: 孙涛 陈思聪 陈江学涯 吴浩 肖婷婷
Sun Tao, Chen Sicong, Chen Jiangxueya,
Wu Hao, Xiao Tingting
U: 吉利控股集团 / 亿咖通科技
GEELY / ECARX

P223 智己 L7 智能驾舱系统（IMOS）
INFORMATION SYSTEM HMI DESIGN
FOR IM L7（IMOS）
A: 李微萌 关超雄 胡青剑 彭璆 方贻刚
Li Weimeng, Guan Chaoxiong, Hu
Qingjian, Peng Qiu, Fang Yigang
U: 智己汽车 / 智己汽车·软件 & 用户触
点团队
IM Motors / IUED

P224 科大讯飞智慧黑板
IFLYTEK SMART BLACKBOARD
A: 雷辰阳 诸臣 许宝月
Lei Chenyang, Zhu Chen, Xu Baoyue
U: 科大讯飞股份有限公司
IFLYTEK Co., Ltd.

P224 智能编程机器车
U-CAR
A: 赵鸿 诸臣 许宝月
Zhao Hong, Zhu Chen, Xu Baoyue
U: 科大讯飞股份有限公司
IFLYTEK Co., Ltd.

P225 网易数字文化中心
NETEASE DIGITAL CULTURE CENTER
A: 林智 袁思思 顾费勇 刘勇成 胡志鹏
Lin Zhi, Yuan Sisi, Gu Feiyong, Liu
Yongcheng, Hu Zhipeng
U: 网易雷火 UX
NetEase ThunderFire UX

P225 BEYOND
A: Zhuoneng Wang, Greg Chen,
Youngryun Cho, Wei-Chieh Wang

P226 (ID)ENTITIES
A: Chukwuma Anagbado, Antonia
Kihara

P226 SOUNDS OF FREEDOM
A: Mutana Wanjira Gakuru, Victor
Ndisya
U: Fiction Entertainment / African
Fiction Academy

P227 联想备授课
LENOVO TEACHING ALL-IN-ONE
A: 沈文京 尹婕 姚涔 安尉 蒙骁
Shen Wenjing, Yin Jie, Yao Cen, An Wei,
Meng Xiao
U: 联想研究院
Lenovo Research

P227 妙笔
MAGIC BRUSH
A: 陈舒窈 张颖 徐浩然
Chen Shuyao, Zhang Ying, Xu Haoran
U: 浙江大学
Zhejiang University

P228 大数据领域全链路数据治理设计重塑
DESIGN RESHAPING OF END-TO-END
BIG DATA GOVERNANCE
A: 陈磊 朱凤瑜 蒋贝妮 胡永棋方
Chen Lei, Zhu Fengyu, Jiang Beini, Hu
Yongqifang
U: 阿里云计算有限公司
Alibaba Cloud Computing Co.,Ltd.

P228 TWINZO
A: Michal Ukropec, Jiri Zila, Michal
Celeng, Patrik Pasko, Tomas Vojtek
U: 5.0 technologies j.s.a.

P229 百度百变人生
BAIDU'S MINI PROGRAM -COLORFUL
LIVES
A: 史玉洁 张勇 刘思任 李爽 陈映钐
Shi Yujie, Zhang Yong, Liu Siren, Li
Shuang, Chen Yingshan
U: 百度在线网络技术（北京）有限公司
Baidu Online Network Technology
(Beijing) Co., Ltd.

P229 小爱同学个人定制智能助手
XIAOAITONGXUE PERSONAL AI
INTELLIGENT ASSISTANT
A: 刘静 贾雪威 伍伩华 薛骁 林兆梅
Liu Jing, Jia Xuewei, Wu Yihua, Xue Xiao,
Lin Zhaomei
U: 北京小米松果电子有限公司
Beijing Pinecone Electronics Co., Ltd.

P230 网易瑶台 —— 会展产业数字化跃迁
NETEASE YAOTAI — ADVANCED
DIGITAL CONFERENCE & EXPO
A: 郭冠敏 刘昊 曾靖喜 曹力文 陈雅萍
Guo Guanmin, Liu Hao, Zeng Jingxi, Cao
Liwen, Chen Yaping
U: 网易伏羲
FuXi

P230 京东智造云 AI 决策仿真平台
AI-BASED DECISION — MAKING
SIMULATION PLATFORM
A: 周澍 迟李青 吕昊 曹铭喆 胡炜
Zhou Shu, Chi Liqing, Lv Hao, Cao
Mingzhe, Hu Wei
U: 京东科技信息技术有限公司
JD Technology Information Technology
Co., Ltd.

P231 格力云数字孪生平台
GREE CLOUD DIGITAL TWIN FACTORY
A: 赖元杰 熊文彬 张法祥 郭正圻 梁根蔚
Lai YuanJie, Xiong Wenbin, Zhang
Faxiang, Guo Zhengqi, Liang Genwei
U: 格力电器股份有限公司
Gree Electric Appliances,Inc.

P231 AKILA DIGITAL TWIN PLATFORM
A: Wilfred Leung Wei Lit
U: Akila

P232 讯飞智能办公本 UI 交互系统
IFLYTEK AINOTE UI INTERACTIVE
SYSTEM
A: 张晗 李守强 程琛 吴蒙勤 肖栋添
Zhang Han, Li Shouqiang, Cheng Chen,
Wu Mengqin, Xiao Dongtian
U: 合肥讯飞读写科技有限公司
iFLYINK

P232 SMART CHILDCARE CENTER /
LOOKMEE
A: Yasuyuki Toki, Akiko Asano, Hiroaki
Akanuma
U: Unifa Inc

P233 A-VIBE
A: Tianqin Lu

P233 与菌绝 - 针对酒店隔离人员的服务设计
SERVICE DESIGN FOR HOTEL
ISOLATION PERSONNEL
A: 陈天易 胡超杰 张琪 刘豪 李熠炫
Chen Tianyi, Hu Chaojie, Zhang Qi, Liu
Hao, Li Yixuan
U: 浙江工业大学
Zhejiang University of Technology

P234 COCOTRUCK
A: Marina Kim, Seunghoon Jeong
U: COCONUT SILO

P234 医疗第三空间
MEDICAL THIRD SPACE
A: 张明鋐 陈玛丽
Zhang Minghong, Chen Mali

P235 斑马数智化导览服务
BANMA INTELLIGENT DIGITAL GUIDER
SERVICE
A: 李经 胡铭洋 余晓瑜 陈思承 张立
Li Jing, Hu Mingyang, Yu Xiaoyu, Chen
Sicheng, Zhang Li
U: 斑马网络技术有限公司
Banma Network Technology Co.,Ltd.

P235 SEEPORT 室内交互型 VR 运动设备
SEEPORT — INDOOR INTERACTIVE VR
SPORTS EQUIPMENT
A: 李玉媚 林安琪 江靓
Li Yumei, Lin Anqi, Jiang Liang

P236 百度知识胶囊
KNOWLEDGE CAPSULES
A: 史玉洁 张勇 韩璐 李浩 欧阳石
Shi Yujie, Zhang Yong, Han Lu, Li Hao,
Ouyang Shi
U: 百度在线网络技术（北京）有限公司
Baidu Online Network Technology
(Beijing) Co., Ltd.

P236 CULTURAL ADVOCATE PROGRAM
A: José Pablo Domínguez, Tako
Sulakvelidze, Claire Yuan Zhuang,
Valentina Palacios, N/A

P237 基于视觉 AI 算法的医疗行为智能分析
与数字化管理系统
MEDICAL BEHAVIOR ANALYSIS
MANAGEMENT SYSTEM
A: 应东东 赵凯 屈世豪 范育芳 董少谦
Ying Dongdong, Zhao Kai, Qu Shihao,
Fan Yufang, Dong Shaoqian

U: 杭州百世伽信息科技有限公司 / 杭州
哲美文化产业发展有限公司
Hangzhou Bestplus Information
Technology Co., Ltd. / Hangzhou Zhemei
Cultural Industry Development Co., Ltd.

P237 商汤绝影智能车舱
SENSEAUTO CABIN
A: 唐御聪 孙康 闫雪 李轲
Tang YuCong, Sun Kang, Yan Xue, Li Ke
U: 北京商汤科技开发有限公司
Sensetime

P238 AED 除颤仪的 AR 培训系统设计
AR TRAINING SYSTEM DESIGN OF AED
DEFIBRILLATOR
A: 王小东
Wang Xiaodong
U: 浙江大学
Zhejiang University

P238 SKOP
A: LECROQ Cyrille, ARNAUD Philippe,
BERNELIN Alexandre
U: WEMED Studio PAD

P239 阿里云数字医疗智能应用
ALIBABA CLOUD DIGITAL HEALTHCARE
INTELLIGENT SYSTEM
A: 廖敏月 许赞 刘漪 张惠顺 叶旭辉
Liao Minyue, Xu Zan, Liu Yi, Zhang
Huishun, Ye Xuhui
U: 阿里云计算有限公司
Alibaba Cloud Computing Co.,Ltd.

P239 华为生态通
ECOCAPTAIN
A: 杨敏敏 唐小敏 陈海武
Yang Minmin, Tang Xiaomin, Chen
Haiwu
U: 华为技术有限公司
Huawei Technologies Co.,Ltd.

P240 NUROX
A: EOJIN JUN

P240 商用清洁机器人及软件平台
COMMERCIAL CLEANING ROBOT AND
SOFTWARE PLATFORM
A: 陆益彬 宋志敏
Lu Yibin, Song Zhimin
U: 美的集团 - 美智纵横科技有限公司
Midea Group

P241 微循环快充充电架
MICRO-CIRCULATION FAST CHARGING
CHARGER
A: 乔楠 刘威 金德智 范志航 程胜超
Qiao Nan, Liu Wei, Jin Dezhi, Fan
Zhihang, Cheng Shengchao
U: 宇通客车股份有限公司 / 郑州飞鱼设
计有限公司
Yutong Bus Co.,Ltd./ Feish Design
（Zhengzhou）Co.,Ltd.

P241 斐视5G 远程驾驶舱
FISON 5G COCKPIT
A: 周才致 田野 张沙 胡圣贤 何遥
Zhou Caizhi, Tian Ye, Zhang Sha, Hu
Shengxian, He Yao
U: 长沙斐视科技有限公司
FISON Technology Co., Ltd.

P242 HERDVISION
A: Matthew Dobbs, Ben McCarthy,
Simon McDonald, Heather Sanders
U: Agsenze Ltd.

P242 COFFEED
A: Zongheng Sun, Yumeng Li
U: LI & SUN DESIGN LLC/PEAR &
MULBERRY

P243 美啊设计平台 —— 培养引领未来设计
师的在线教育平台
MEIA DESIGN PLATFORM— AN
ONLINE EDUCATION PLATFORM
A: 胡晓 苏菁 胡蓉 罗志国
Hu Xiao, Su Jing, Hu Rong, Luo Zhiguo
U: 广州美啊教育有限公司
Guangzhou Meia Education Co. , Ltd.

P243 VUVU 生物教具
VUVU
A: 杨俊辉 卓宜萱 简邱伟 许品宜 陈沛颐
Yang Junhui, Zhuo Yixuan, Jian Qiuwei,
Xu Pinyi, Chen Peiyi
U: NetEase ThunderFire UX

P244 ZHY963 —— 智慧融合可视化终端
ZHY963 — INTELLIGENT FUSION
VISUAL TERMINAL
A: 赵砚青 刘庆超 范仕亮 王明立 杨帆
Zhao Yanqing, Liu Qingchao,Fan
Shiliang, Wang Mingli, Yang Fan
U: 智洋创新科技股份有限公司
Zhiyang Innovation Technology Co., Ltd.

P244 第四范式 AIOT—— 智慧园区综合决策系统
AIOT — SMART PARK INTEGRATED
DECISION-MAKING SYSTEM
A: 迟娩 王凯 徐昀 王蔚
Chi Mian, Wang Kai, Xu Yun, Wang Wei
U: 第四范式（北京）技术有限公司
Fourth Paradigm (Beijing) Data &
Technology Co., Ltd.

P245 CIRCLE SAVINGS
A: Patrick Awori
U: Imaginarium

P245 城市与我 —— 创新型城市数字孪生系统
CITY&ME — INNOVATIVE DIGITAL TWIN
SYSTEM OF CITY
A: 詹明明 王思越 苗雨菲 陈俊锰 黎咸训
Zhan Mingming, Wang Siyue, Miao Yufei,
Chen Junmeng, Li Xianxun
U: 中国电子系统技术有限公司
Chinese Electronic System Technology
Co., Ltd.

P246 腾讯技术公益志愿者平台
T-VOLUNTEER NETWORK
A: 邱颖彤 郑露 张翘 陈晓畅 饶瑞
Qiu Yingtong, Zheng Lu, Zhang Qiao,
Chen Xiaochang, Rao Rui
U: 腾讯科技（深圳）有限公司
Tencent Technology (Shenzhen) Co.,Ltd.

P246 智能矿山
INTELLIGENT MINING
A: 高健 杨敏敏 文梦芝 柳玮 朱磊
Gao Jian, Yang Minmin, Wen Mengzhi,
Liu Wei, Zhu Lei
U: 华为技术有限公司
Huawei Technologies Co.,Ltd.

P247 BOTTO! — THE IOT DEVICE AND
CHATBOT AGAINST FOOD WASTE
OPENDOT
A: Enrico Bassi, Vittorio Cuculo, Antonio
Garosi
U: OpenDot

P247 PLANTIVERSE
A: Cecilia Tham, Mark Bunger, Magda
Mojsiejuk, Nuria Albo,David Tena
U: Futurity Systems

图书在版编目（CIP）数据

2022中国设计智造大奖年鉴 = 2022 DESIGN
INTELLIGENCE AWARD YEARBOOK / 中国设计智造大奖组委
会编. — 北京：中国建筑工业出版社, 2023.12
　　ISBN 978 - 7 - 112 - 29353 - 7

　Ⅰ. ①2… Ⅱ. ①中… Ⅲ. ①工业设计 — 中国 —
2022 — 年鉴 Ⅳ. ①TB47 - 54

中国国家版本馆CIP数据核字(2023)第215810号

责任编辑：吴绫 杨晓
责任校对：王烨

2022 中国设计智造大奖年鉴
2022 DESIGN INTELLIGENCE AWARD YEARBOOK

中国设计智造大奖组委会 编

*

中国建筑工业出版社 出版、发行（北京海淀三里河路9号）
各地新华书店、建筑书店经销
当纳利（广东）印务有限公司印刷

*

开本：889 毫米 ×1194 毫米　1/12　印张：25 ⅓　字数：732 千字
2023 年12月第一版　　2023 年12月第一次印刷
定价：338.00 元
ISBN 978 - 7 - 112 - 29353 - 7
(42089)